QE
41
B67

Boyer, Robert Ernst, 1929–
    Activities and demonstrations for earth science, by
Robert E. Boyer and Jon L. Higgins. West Nyack, N. Y.,
Parker Pub. Co. ₁1970₁

    286 p. illus. 25 cm.

    Bibliography : p. 268–282.

    1. Earth sciences—Experiments.   ɪ. Higgins, Jon L., joint au-
thor. ɪɪ. Title.

QE41.B67   550′.28   72–104721
SBN 13–003582–3   MARC

Library of Congress   70 ₁12₁

Activities and Demonstrations
for Earth Science

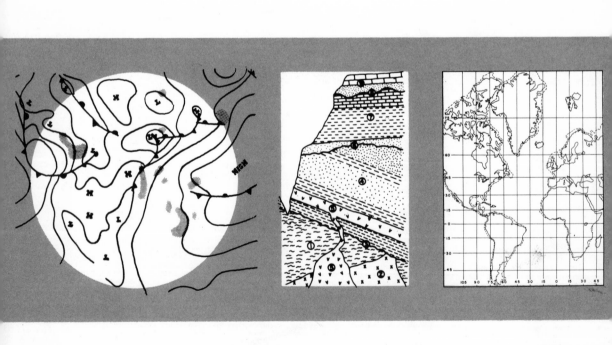

ACTIVITIES AND DEMONSTRATIONS FOR EARTH SCIENCE

BY

ROBERT E. BOYER

Professor of Geology and Education

The University of Texas at Austin

AND

JON L. HIGGINS

Assistant Professor of Education

Stanford University

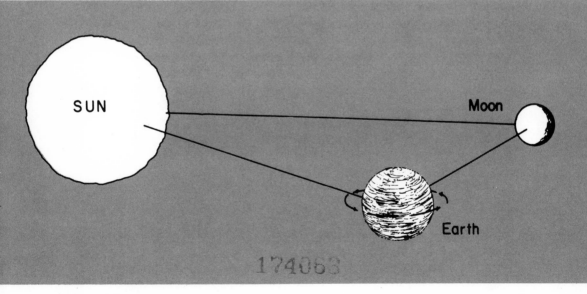

Parker Publishing Co., Inc.   West Nyack, N.Y.

PRINTED IN THE UNITED STATES OF AMERICA
13—003582—3 B & P

Adapting Investigations to an Earth Science Course

This book of inquiry experiences is intended as a teaching aid for junior and senior high school earth science teachers. The experiments it contains are open ended and easily adapted to use at many different grade levels and with any textbook or subject matter approach. They can be used in qualitative ways to give younger students insight into the nature and relative importance of the concepts and processes of earth science. They can be equally effective for giving more experienced students the opportunity to make detailed observations, measurements, and analyses of earth science. We have made an effort to incorporate appropriate degrees of freedom in each experiment so that students can devise their own specific procedures according to their interests and abilities.

As a help to you, the earth science teacher, each of the basic investigations in the book is organized and presented for ready use. Each includes motivating suggestions, background information, lists of materials, student experimental procedures, analysis of results, and ideas for follow-up investigations. Pertinent source information is also provided. For easiest use, experimental procedures are in step form and in language directed to the student. Necessary materials and apparatus are generally simple and inexpensive.

The follow-up investigations should be especially useful with students at the upper grade levels (10-12) and highly motivated students in the lower grades (7-9) who are interested in those particular subjects. A few of the additional experiments require field trips for firsthand observations of earth processes. Several are extended investigations which require students to make repeated observations over an interval of two or three weeks before they can reach meaningful conclusions.

5

Teachers of students who have less than a full academic-year course in earth science can make a selection of some of the experiments in the book to serve as illustrations of the concepts involved. In any case, we recommend that whenever possible the experiments be conducted as classroom investigations with students participating. When this is not possible, however, the experiments can be given as demonstrations by the teacher.

We have tried to embody some aspects of the working scientist's viewpoint and practice in this book. Scientists almost never experiment to check and confirm something they already know. They perform experiments to explore hunches and investigate possibilities, and we have tried to reflect this philosophy in the book. The most valuable experiments are not necessarily those yielding definite facts and final conclusions. Commonly, the most valuable experiments are those that uncover new questions, suggest new investigations, and open up whole new lines of inquiry. We have attempted to embody this spirit of scientific inquiry in the book.

Few of the experiments in this book lead to a definite closure. Instead, most of them provide reasons for doing the following experiment. Since the experiments take on deeper meaning in relationship to each other, it is appropriate to think of them as experiment sequences. The sequencing of these experiments follows a logical thought process within the disciplines in earth science—geology, oceanography, meteorology, and astronomy. We have followed this organization primarily for teaching convenience. However, the teacher should be continually aware that many earth processes extend across these arbitrary subdivisions, as illustrated by the hydrologic cycle, which involves the continents, oceans, and atmosphere and is driven by energy from the sun. It is important to view each of the experiments across subdivisions as they relate to major concepts and processes in earth science.

Seeing beyond surface facts to underlying processes only begins with investigative experiences like those presented here. It must ultimately be nurtured by the addition of textbooks and by the teacher's insight. We hope that this book of experiments will provide experiences which will give your students a better understanding of some of the ways in which a scientist works and thinks. In addition, we hope that some of the questions raised by the experiments will lead you to further reading and, subsequently, to further open-ended activity in teaching earth science.

ROBERT E. BOYER

AND

JON L. HIGGINS

Acknowledgments

This series of experiments has evolved through three years of writing and testing. We are indebted to the many teachers and students whose comments and feedback molded and improved these activities. We especially thank Miss Kathora Remy, Consultant for Science, San Antonio Independent School District, who worked with us on a Cooperative College-School Science Program (NSF Grant GW-708) when many of these experiments were tested and coordinated into a usable classroom sequence. We are grateful to Mr. Lawrence J. Buford, Supervisor of Secondary School Sciences, Austin Independent School District, who cooperated with us in further testing several experiments in an In-Service Institute in Earth Science for Secondary School Teachers (NSF Grant GW-2094).

Several activities are modified from experiments developed in curriculum programs with which the authors have been affiliated: the Earth Science Curriculum Project (R.E.B.) and the Illinois Astronomy Project (J.L.H.). Prof. W. H. Matthews III, Lamar State College of Technology, contributed to the activities treating fossil studies; Mr. Wayne R. Schade, Menomonee Falls East High School, Menomonee Falls, Wisconsin, worked on several experiments in oceanography; and Prof. Lynton S. Land, The University of Texas at Austin, aided in developing the activity on water circulation in the oceanography section.

We are also indebted to Mr. Richard B. Hale, Murchison Junior High School in Austin, who critically reviewed the manuscript and made many suggestions for its improvement.

Mr. Donald P. Erickson drafted the illustrations; the authors did the photography, except as noted.

R.E.B. and J.L.H.

7

Contents

CONTENTS

Section One

GEOLOGY

1

How Big Is the Earth?

MOTIVATORS

Let's find out how big the earth really is! To do this, we first need to measure how far it is around the earth (the earth's *circumference*). Of course, we must assume that the earth is a sphere. Ask the students to list how many different ways they can think of to measure the circumference of the earth.

BACKGROUND INFORMATION

Over 2000 years ago a Greek geographer named Eratosthenes became aware of a deep well near the city of Syene, now called Aswan, in southern Egypt, which exhibited an interesting feature. Once each year, at high noon, the sun shone directly down the well, clearly lighting the bottom. Eratosthenes reasoned that the sun was then directly over the well—and thus the light rays fell perpendicular to the surface. However, at the same time in the city of Alexandria, to the north of Aswan, a tall, vertical obelisk (four-sided, tapering stone pillar shaped at the top like a pyramid) cast a shadow. Thus the sun's rays were not perpendicular to the surface at Alexandria.

Eratosthenes also knew the distance between Aswan and Alexandria. In his time, distances were stepped off by men called "pacers," who were trained to walk so that each step covered the same distance. Pacers walked from city to city, counting each step they made. The distance from Aswan to Alexandria was paced off at approximately 925 kilometers.

Eratosthenes assumed that rays of sunlight reaching earth travel in essentially parallel paths. [This is correct because the sun is so far away that its light comes in parallel rays. No matter where we stand on earth, if everyone points at the sun at the same moment, they point in parallel directions (Figure G-1). This means that light rays from the sun to earth never meet or cross.]

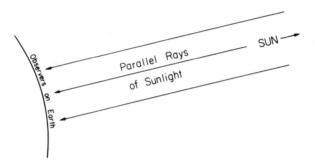

Figure G-1. *Rays of sunlight are shown approaching earth in parallel paths.*

Eratosthenes also believed that the earth was a sphere. He then decided to measure the earth's circumference. To do this, he measured the angle of a shadow cast by the obelisk in Alexandria at the same time that the sun's rays reached the bottom of the well in Aswan. As shown in Figure G-2, Eratosthenes knew that this angle was equal to the angle at the earth's center between Aswan and Alexandria (he understood geometry which shows that these angles are equal because the rays are parallel).

The angle he measured was slightly over 7°. Because there are 360° in a circle, Eratosthenes figured that this angle was about $\frac{1}{50}$ of a circle.

$$\frac{7° \text{ (measured)}}{360° \text{ (in circle)}} = \text{approximately } 1/50$$

The distance on the surface between these two cities (925 km) was thus approximately $\frac{1}{50}$ of the total distance around the world.

16

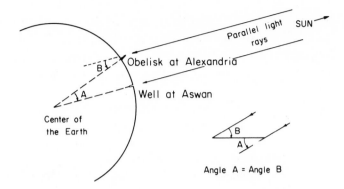

Figure G-2. *Method used by Eratosthenes to determine circumference of the earth.*

Hence by multiplying (925 km x 50), he figured that the circumference of the earth was approximately 46,250 km. This was an amazingly good estimate when his method is compared to modern techniques for making these measurements. (The accepted value for the earth's circumference measured at the equator is 40,076 km.)

MATERIAL

- Styrofoam ball (preferably 10 cm or larger in diameter)
- 2 toothpicks
- String (approximately 30 cm long)
- Ruler
- Protractor

STUDENT PROCEDURE

1. We can apply Eratosthenes' technique to measure the circumference of a sphere. Take the styrofoam ball and place two toothpicks into it. Space the toothpicks 4 to 6 cm apart. Be sure to insert the toothpicks so that they point *toward the center* of the ball. Allow them to stick up about 2 cm.
2. Hold the ball close to a light source so that the light rays shine directly on one toothpick and no shadow is cast by it. (This is analogous to the well at Aswan.) If convenient, use the sun as the light source;

17

however, if weather or other factors prevent use of direct sunlight, a lamp will be entirely adequate.

3. Holding the ball in this position, have a classmate mark the length of the shadow from the other toothpick (analogous to the obelisk at Alexandria).

4. Now make the following measurements:

   a. *The angle between the toothpick (casting shadow) and a line from the top of the toothpick to the end of the shadow.* (This is a little hard to measure. Hold a piece of paper up to the ball and trace the length of the toothpick and mark the end of the shadow. Then draw the line from toothpick top to shadow end and use your protractor to measure the angle at the top of the toothpick.)

   b. *The distance between the toothpicks along the ball.* (If your ruler bends, you can follow it along the ball. Otherwise, use a piece of string to mark the distance between the toothpicks and then measure this length of string.)

5. You now have the information for a sphere, comparable to the data Eratosthenes obtained for the earth in order to determine its circumference. The distance measured between the toothpicks is a distance of arc on the ball and the angle measured is equal to the angle of this arc at the center of the ball. With this information, you can figure the circumference of the ball realizing that the angle at the center of the ball would then be 360°. *(Hint:* You may want to figure the distance of arc that would represent an angle of only 1° so that you can multiply by 360.)

ANALYZING RESULTS

After you finish (not before), wrap a piece of string around the ball and then measure its length. How does it compare with the calculated circumference? If the two answers are different, explain why this could happen.

Eratosthenes thought that Alexandria was due north of Aswan. Thus a plane through Alexandria, Aswan, and the center of the earth would also include the noon sun. However, Aswan is not due south of Alexandria and this caused some error in his measurements.

INVESTIGATING FURTHER

By following the procedure used by Eratosthenes, students determined the circumference of a sphere. They can now check the earth's circumference using similar information. Milwaukee, Wisconsin is approximately 10°, nearly due north of Tuscaloosa, Alabama. The cities are about 1115 km apart. Using these figures, have students calculate the earth's circumference measured around the poles. How does this compare with the accepted polar circumference of 40,008 km?

By careful measurements with sensitive instruments, scientists have been able to prove that the earth is not a perfect sphere. It bulges slightly around and south of the equator so that it has somewhat of a pear shape. (The equatorial circumference is about 68 km greater than the polar circumference.)

How have photographs and measurements from satellites helped to prove that the earth is essentially spherical? Even with this evidence there is still a group of people (in London, England) who call themselves the Flat Earth Society!

2

Detecting Geologic
History Through
Fossils

Ask the students what they think the earth was like millions of years ago. Do geologists really know what the condition of the earth's surface was that long ago? How might we find out about the history of the earth? Fossils are one important clue—they can give us a view into the nature of the earth in the geologic past. Not only can we tell the kinds of plants and animals that once lived here, but by studying fossils we can learn much about the environment in which this past life existed. How do we go about studying fossils so that they reveal these secrets to us?

BACKGROUND INFORMATION

A *fossil* is any evidence of life in the past. Many fossils are the actual remains of hard parts of ancient organisms, but some are just evidences of former life such as footprints and carbon residues. Fossils serve as one of the major means of affording information about the geologic past—the type of life that existed, the environments in which the organisms lived, the assemblages of associated plants and animals, the completeness of the depositional record as evidenced by sedimentary rocks, and the understanding of processes in sedimentation (depth of water, rate of sediment deposition, salinity of water, etc.) that led from deposition of unconsolidated sediments to the eventual consolidated sedimentary rocks.

The variety of types of fossil preservation include:

(1) Complete remains—An example is woolly mammoths (extinct variety of elephant that lived in cold northern climates) frozen in glaciers since the Ice Age. Some of these were so well preserved that the meat was reportedly fresh enough to eat when the carcasses were uncovered from the melting ice.

(2) Unaltered hard parts—Particularly teeth, bone, shell, and wood materials.

(3) Petrified remains—Original material replaced by mineral matter (especially calcium carbonate or silicon dioxide replacement of wood and bone), with preservation of the internal textural detail.

(4) Molds and casts—When original organic material is removed from rock, the hollow space remaining is a *mold*. If this mold is subsequently filled, the filling is a *cast*. Molds and casts reflect only the shape and surface features of the organisms, and not the internal structure.

(5) Carbon residues—A carbon-rich film of an organism, especially common for some leaves and fish.

(6) Tracks, trails, and burrows—Prints and burrow marks without any actual remains of the organism.

(7) Indirect evidences—Stomach stones (*gastroliths*) that were swallowed by some types of animals and used in their digestive process. Excrement (*coprolites*) preserved by petrification that give clues to the diet of the animals.

MATERIAL

- Footprint diagram—preferably as a transparency if overhead projector is available (Figure G-3, next page).

PROCEDURE

Footprints can tell us much about the nature of the animals that made them. From a single footprint, paleontologists (scientists who study plant and animal life of the past) may be able to learn the general size and

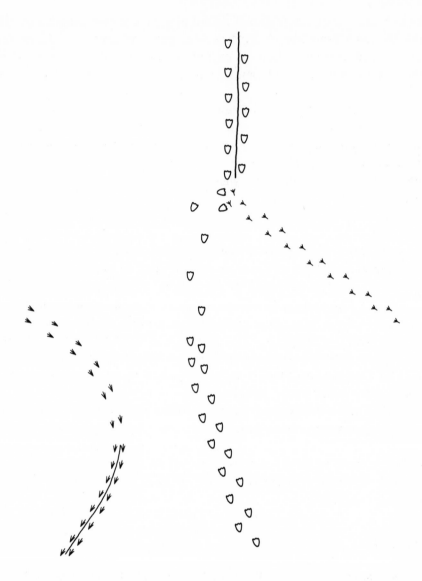

Figure G-3. *Pattern of fossil tracks that "tells a story" about the dinosaurs which made them.*

nature of the animal that made it, as well as learning about the environment in which the animal lived. Sets of tracks may, therefore, be exceedingly informative:

    1. Ask questions such as the following to start the class thinking about footprints.

    a. Where could a person or animal walk today to leave footprints that might be preserved as future fossil prints?

    b. How can you tell the direction in which an animal walked from footprints? What if the animal walked backwards?

    c. How will the footprints of an animal differ when it walks and then when it runs?

2. Place a transparency copy of the footprint puzzle (Figure G-3) on an overhead projector. Begin with the entire transparency covered with a piece of cardboard and gradually uncover it by sliding the cardboard off from bottom to top. As the transparency is uncovered, a little at a time, ask the students to infer what is happening. Encourage ideas from the students with questions such as the following:

    a. What is the line between the footprints of the smaller animal? Why does this line disappear when the footprints are farther apart? (It could be something the animal was dragging in its mouth and dropped when it began to run. Or perhaps the animal was dragging its tail when it walked but when it ran its tail was lifted up.)

    b. Which way do you think the wind was blowing at the time these footprints were made? (It appears that the first of the smaller animals sensed the presence of the larger animal, but the second smaller arimal walked right up to the larger animal. So, the wind probably was blowing from right to left.)

    c. How many legs are represented by each animal? (Probably only two. Many carnivorous dinosaurs had short front limbs and walked on their hind legs.)

    d. Can you tell when these animals changed their speed? (This is readily seen by the distance between the footprints.)

    e. What was the probable nature of the land surface when these footprints were made? (This probably took place on a moist, soft soil because of the well-preserved tracks. Perhaps on a beach or even in a shallow sea.)

ANALYZING RESULTS

A lesson in scientific thought should be brought to the students' attention. Encourage them to make inferences and formulate hypotheses based on the data presented as the footprint puzzle is unveiled. Do not stifle imaginations, but caution the students against making interpretations to fit a preconceived story. They should become aware of the necessity to test their hypotheses continually in light of additional evidence (in this case, seeing more of the puzzle unveiled), and be ready to alter their views if the new evidence contradicts previous ideas. A good scientist admits when he is wrong, and is quick to discard an hypothesis proven wrong.

It is permissible (even desirable) to conclude with alternative hypotheses if they appear acceptable. Perhaps the meeting of the large and small animals resulted in death to the smaller one. The heavy line between the footprints of the larger animal suggest that it is dragging the smaller animal away. But it is possible to conclude that this line results from the larger animal dragging its tail. The smaller animal could have been an offspring, being met and carried away by its mother.

In addition to "telling a story," the fossil tracks yield information about the environment of the area during the time when the tracks were formed. From this information, the paleontologist can obtain a more complete picture of the geologic history of the region. Features such as land-sea distribution, general climatic conditions, and the nature and proximity of food sources (grasses, plants, and trees for the herbivorous dinosaurs) can be discerned with a reasonable degree of accuracy. By comparison of the bits of information obtained at this fossil locality with studies of rocks (and the fossils they contain) in other areas, a general picture through geologic time can gradually be assembled. Although it may be slow, painstaking work, it commonly results in a detailed geologic history of the region.

INVESTIGATING FURTHER

After the discussion, ask the students to prepare footprint puzzles. Emphasis should be placed on normal situations with due consideration of the natural environmental conditions. The best of these puzzles can then be presented in class for analysis.

Some classes may wish to learn about fossil footprints by making new footprints. This requires containers (of fairly large area) with a layer of damp clay. Have different students step on the clay and determine the relationship between the depth and size of footprint and the weight and height of each person. Results can be graphed to see if the plot is consistent.

The importance of this approach is the evaluation of the experimental method. Perhaps the graphed results reveal some differences that appear anomalous. Two variables have been involved here, weight of person and area of the footprint. The area over which the weight is spread affects the depth of the print—the smaller the area, the deeper the print. In order to determine relationships, it is necessary to change only one thing at a time and observe the results. A heavy cardboard footprint could be made and have each student stand upon this print on the clay. Then the only variable is the weight of the student. How many variables are involved in the actual making of fossil footprints?

3

Characteristics of a
Fossil Population

MOTIVATORS

Extensive study of fossils—and a comparison of them with living organisms—first led Charles Darwin to propose a theory of change in life forms through evolution over 100 years ago. Although even to this day some controversy of the theory of evolution still exists, evidence in support of Darwin's concept is mounting. A study of fossils (collected locally if possible) will provide students with some information to appreciate variations in species and perhaps furnish some observations pertinent to the theory of evolution.

BACKGROUND INFORMATION

According to Darwin's theory, changes in life forms take place because individuals best suited to their environment most commonly survive, and, in turn, pass on their desirable characteristics to their offspring. He termed this *natural selection* because nature (the competitive environment in which the organisms live) in essence does the selecting. Those individuals who find difficulty competing (primarily for food and protection against predators) may die or migrate to an area where they are bet-

26

ter adapted for competing. Thus the members of one species may tend to become very similar as they all acquire the same desirable characteristics for survival—especially within one local environment.

MATERIAL

- Approximately 24 specimens of a species of fossil per student group
- Graph paper
- Measuring rule

STUDENT PROCEDURE

Ideally, the fossils used for this investigation should be collected by the students on a field trip. If the local area does not afford fossil localities, or if other factors preclude such collecting, fossils obtained in other ways are quite acceptable. The investigation is most effective if the students work in small groups (not to exceed four per group).

1. Decide what two features you wish to measure for each sample, for example: length and width, length and height, width and number of coils, etc. In one study the maximum width and length were measured for the Cretaceous oyster, *Exogyra texana* (pictured in Figure G-4).

Figure G-4. *Samples of the Cretaceous oyster,* Exogyra texana, *used for study of the characteristics of a fossil population. Scale shown by bar which is 1 cm long. (Photograph courtesy of Gale A. Bishop.)*

27

Obviously, some care must be taken to select a particular fossil species with a shape that has two features that can be measured conveniently.

2. Then make the same two measurements on each specimen in the fossil set.[1]

3. Graph the results on a coordinate system as shown in Figure G-5.

ANALYZING RESULTS

Normally, the graphed measurements will plot in a reasonably well-defined straight line as shown in Figure G-5. This implies a relatively uniform correlation of the changes between specimens of different sizes. (The sizes represent different growth stages of the same species, not different species.) If all the fossils are virtually of the same size, the plotted points will concentrate in a tight cluster.

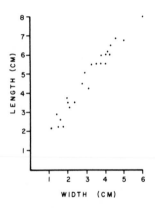

Figure G-5. *Graph showing plot of length and width of 24 specimens of the fossil,* Exogyra texana.

Any significant departure from the general plotted pattern should be discussed. Perhaps the anomaly represents a specimen of a different species which was inadvertently included, or perhaps the specimen was deformed (e.g. flattened by compression) during burial.

If sets of reproduced fossils as described in Appendix I (Method 1) are used, graphed results of the different student groups should be essen-

---

[1] Appendix I (Method 1) outlines an inexpensive and efficient procedure for reproducing fossils. By using this technique, an indefinite number of sets of specimens can be reproduced from one set of original fossils. This method can be utilized effectively for those who are able to obtain only one or a limited number of sets of fossils.

tially identical. This affords a good check on the procedure of each group. Also, under these circumstances, a "ringer" or two—a deformed specimen or one of a slightly different-appearing species—can be purposely included in each set. These would plot outside the desired limits and thus prompt the discussion outlined above.

By using different sets of fossils (either real or reproduced), a discussion of the comparison of results can be conducted. Similar fossils of other ages (or found at other localities) can then be compared with those measured and the significance of these differences discussed. In this manner, species variations can be recognized. Perhaps the observed changes reflect modifications of the animals for better survival. If the environment alters slowly enough, perhaps over hundreds of years, the animals are able to modify their forms continually and thus continually survive. If, on the other hand, the environment shifts drastically or over a span of only a few years, the species may become extinct. The importance of variations noted with respect to the number of specimens measured should also be raised. Was 24 specimens a sufficient number or would significantly different results be obtained by measuring 12 more specimens?

INVESTIGATING FURTHER

Considerable additional insight into molds and casts can also be obtained by discussing the procedure for reproducing fossils described in Method 2, Appendix I. This procedure can be done effectively in class or by individual students at home. The reproduced molds and casts can be compared with real fossils to learn more about the limitations in this kind of fossil preservation. The natural processes that result in the formation of molds and casts should then be examined. Why aren't all molds filled? (Perhaps they were once filled with softer material which has been removed by weathering and erosion of ground water.)

Comparison of the fossils with present-day forms (either living specimens or specimens which have been preserved) can lead to a greater understanding of the bases for Darwin's theory of evolution. Comparing changes that can be observed and evaluating the reasons for these changes are important. Some changes that have taken place are in the soft parts of the organisms and thus cannot be readily detected by study of fossils. However, changes involving features such as general size, ornateness, addition or deletion of appendages, etc. can be evaluated. Be sure to consider the major factors of food-getting ability and protection from predators when discussing reasons for these observed changes.

4

Minerals–The Ingredients
of Rocks

MOTIVATORS

We have already seen that by studying fossils contained in rocks, we can get clues about the earth's history. But can we also learn of past events by studying the rocks themselves? Are the layers of sedimentary rocks like "pages of a history book of the earth"? Three distinct types of rocks exist (See Geology Experiment 5, "Classifying Rocks") and these rocks contain a variety of different minerals. A study of minerals is important not only for the identification of the various rock types, but also to understand the origins of these rocks.

BACKGROUND INFORMATION

Rocks are composed of aggregates of minerals, and minerals in turn are made up of atoms. The nature of these atoms—their size, electrical charge, and arrangement within the minerals—controls the character of the minerals which they form. (Refer to Appendix II for information on the nature of minerals.)[1]

---

[1] Appendix II also outlines the common physical properties of minerals that can be readily tested. By examining unknown specimens for these properties, a majority of the more common minerals can be quickly identified.

Because each mineral is composed of different combinations and arrangements of various atoms (called its molecular structure), each mineral responds to exposure at the earth's surface in a different way. Many minerals react with the atmosphere and ground water to change to new minerals. Minerals such as biotite mica, calcite, halite, and pyrite may be susceptible to chemical or physical change in the relatively short time of only a few years' exposure. They are, therefore, not commonly found except where freshly dug, for example in quarrying, mining, and so forth. In contrast, a few minerals commonly occurring in rocks show little or no reaction to surface exposure, even after many years, and maintain their form very well. These minerals are therefore easiest to find as they "persist" at the earth's surface, whereas the less stable minerals alter to form loose, fine rock debris and soil.

Quartz is the best example of a very stable common mineral. It is, therefore, a prominent constituent of most sands (fine-grained, loose rock and mineral fragments) commonly found on beaches, along the banks of rivers and lakes, and wherever water or wind have caused the accumulation of rock material.

MATERIAL

- Sample of sand and/or gravel

- Collection of minerals (if available)

- Materials for testing physical properties to include:
  streak plate, penny, dull knife (or nail as substitute), small glass plate, magnet, container of dilute hydrochloric acid (about 10 to 15% solution), balance, and graduated cylinder.

- Hand lens or magnifying glass

STUDENT PROCEDURE

Special mineral collections are not needed for this investigation although they can be utilized if available. The sand and/or gravel will suffice nicely and can be collected by the class if convenient.

1. Individually or in small groups examine the mineral material and prepare a list of features which can be used to recognize the different minerals.
2. Discuss these ideas, writing the most useful ones on the blackboard. (Generally this list includes the most obvious features such as color, shape, weight, etc.)

3. *After* having considered how to recognize the minerals, compare these features with the physical properties listed in Appendix II. Go through the order of tests of the physical properties as outlined in Appendix II for each of the different minerals in the collections.

4. Tabulate the results of these tests for each sample (see Table G-1). Samples may contain more than one mineral—this is especially true if gravel is examined. Be sure to test the properties of one mineral at a time. If you are not sure, the sample should be discarded.

5. Based on the results of the tests, try to identify each sample by mineral name using Appendix II and references on minerals available in your library (see Appendix V).

Table G-1. *Tabulation of physical properties for mineral specimens.*

Physical Properties	Specimen #1	Specimen #2	Specimen #3
Luster			
Hardness			
Color			
Streak			
Transparency			
Crystal Form			
Break Cleavage Fracture			
Specific Gravity			
Reaction to Dilute HCl			
Special Characteristics			

ANALYZING RESULTS

A comparison of the results of the tests reveals that each group of students had some of the same minerals. Quartz is undoubtedly abundant. Other common minerals might be biotite mica, calcite, feldspar, hematite, and pyrite. However, the importance of this investigation, and thus the emphasis of the discussion, is not identifying the minerals by name. This

32

should be secondary. The important thing is realizing that unknown specimens can be tested in a systematic fashion because each mineral has distinctive physical and chemical characteristics. Emphasize that systematic testing is a scientific procedure applied to unknown specimens by mineralogists.

One important reason for identifying minerals is because they are the constituents of rocks. By recognizing what minerals they contain, the identity of rocks can be learned more readily. Identifying rocks is necessary in order to understand their origin, and thus more about the history of the earth as shown in the next investigation.

INVESTIGATING FURTHER

Challenge the ability of interested students by giving them a variety of minerals including some less common kinds. Minerals displaying good crystal form can usually be obtained from a local rock shop or from one of the supply companies which sells rocks and minerals. Generally a modest collection of different mineral specimens including some with attractive crystal form can be acquired at little cost. After identification, these collections serve as excellent display specimens.

Students should be encouraged to collect samples of the different minerals on their own by taking local field trips. They should also use available books about minerals in making the identifications. Some books have detailed descriptions of the characteristics of minerals and many contain plates of colored photographs which aid in verifying identifications.

More advanced groups can be assigned to prepare models of the different crystal forms (refer to the discussion of crystal systems in Appendix II). These models serve as an aid in visualizing the written descriptions of crystal form. Actual minerals displaying good crystal form can be compared with these models. The students can be challenged to find at least one mineral displaying the same crystal habit as each of the models.

33

5

Classifying Rocks

MOTIVATORS

Rocks are aggregates of minerals; therefore, the first step to understanding the type and origin of a rock is to identify its mineral constituents if possible. But many of the same minerals occur in different kinds of rocks. Rocks of entirely different origins can even have identical mineral compositions! Obviously, therefore, mineral composition alone may be insufficient to distinguish the origin of an unknown rock sample. What other features might be observed to give the needed information?

BACKGROUND INFORMATION

The *texture* of a rock—described as the size, shape, and arrangement of the constituent particles—may be vital to an understanding of the rock's origin. Texture is controlled by the way a rock forms—that is, its origin. *All rocks can be classified into one of three types on the basis of origin.*

Igneous rocks form by cooling from a molten state, either at some depth within the earth (*plutonic* igneous rock) or at the earth's surface (*volcanic* igneous rock). During the cooling process, the minerals grow as interlocking crystals. This crystalline texture characterizes igneous rocks, and the rate of cooling, to a large degree, controls the general size of the crystals. Thus volcanic rocks, being at the surface, will cool rapidly and contain many small crystals, whereas plutonic rocks cool slowly (losing heat slowly) and are characterized by large crystals.

Sedimentary rocks form by the accumulation of loose rock material (called sediment) which is compacted and cemented into hard rock. Sediments may be various-sized particles derived from rocks which occur at the earth's surface, such as mud, sand, and gravel, or material accumulated by precipitation from solution out of a body of water (primarily the oceans). Some sediments contain organically derived material, including fossils. Sedimentary rocks are characterized by layers (called beds or strata) which result by hardening of the layers of sediments originally deposited. They may also contain fossils which are unique to sedimentary rocks. Some sedimentary rocks can be distinguished by the presence of sediments as grains which have been cemented together.

Metamorphic rocks result from changes to pre-existing rock (which may formerly have been either igneous or sedimentary) due to increased pressure and temperature. These conditions cause the minerals to recrystallize into new or larger crystals. Because pressures are generally not uniformly applied, many metamorphic rocks contain a pronounced orientation of mineral shapes forming a banded or platy-mineral sheeted texture known as foliation or schistosity.

Some problems may be encountered during first attempts to group rock samples into these three categories. However, with some practice and reference to available literature (refer to Appendix III),[1] this identification should be accomplished without difficulty.

MATERIAL

- Collection of rocks
- Hand lens or magnifying glass

STUDENT PROCEDURE

Special rock collections are not necessary for this investigation. How-

---

[1] Appendix III lists the features most readily apparent for identification of some common rocks in each of the three types of rocks.

ever, it is desirable to have representative samples of each of the three types of rocks (preferably at least two different samples representing each origin).

1. Individually or in small groups, examine the rock collections and prepare a list of the features you think are useful to recognize the three main types of rocks and to name the rocks within each type.

2. Discuss these lists of features, writing the most useful ones on the blackboard.

3. *After* having thought about the problems in recognizing rocks, study the data sheets that are handed out (Table G-2).

4. Using available references and Appendix III, determine the features given on the data sheet for each rock sample.

5. Based on these observations, identify the rock type and give a rock name to each sample.

Table G-2. *Tabulation of features in rock samples to be studied.*

Features of the Rock	Sample #1	Sample #2	Sample #3
1) Size of constituent particles			
2) Shape of constituent particles			
3) Identification of constituent particles (crystals, grains, rock fragments)			
4) Minerals identified a. b. etc.			
5) Layering (bedding), metamorphic banding or foliation, etc.			
6) Fossils			
7) Rock type (igneous, sedimentary, metamorphic)			
8) Rock name (sandstone, granite, schist, etc.)			

ANALYZING RESULTS

The identity of some samples will be quite obvious to the students, whereas others will be of concern. As with minerals, it is impossible to set up a foolproof system leading to the correct identification of every sample. The discussion should then focus on other information which might help to identify the "mystery" sample. The geographic source of the rock is of particular importance! Although this can be misleading, it is usually a significant clue to correct identity of the origin. (This is true because it is common to find rocks of one type, for example volcanic igneous rocks, concentrated in specific geographic areas.)

Be sure to compare known minerals (identified in the previous experiment, "Minerals—The Ingredients of Rocks"), with these same minerals (identified or thought to be present) in the rock samples. Commonly, this will confirm the tentative rock identification as correct or will suggest that it is in error.

Do not be dismayed if students fail to identify correctly all the rocks. The differences between some rocks are actually very difficult to recognize. Some metamorphic rocks, for example, may have properties similar to the "parent" igneous or sedimentary rock from which they were derived. And some rocks are composed of such very fine particles that it is difficult to tell whether they are tiny interlocking crystals or fine grains cemented together—even when using a hand lens or magnifying glass.

INVESTIGATING FURTHER

Caution the students to realize that the classification of rocks is man made. Nature doesn't always conform to this classification. Can you think of examples where this is especially true for rocks? Have the class list possible conditions which would result in the formation of rocks with "mixed" features. (Included might be volcanic ash which settles in a lake and thus becomes sediment, very slightly metamorphosed shale that still contains some fossil fragments which can be recognized, weathered igneous rock that looks like sedimentary rock, and rocks found in the border zone between plutonic igneous rock and the metamorphosed rock that surrounded it.)

On the basis of their understanding of the origins of rocks, have the class list all the possible things that can happen to rocks as they change from one type to another—weathering, erosion and rounding, heating, melting, deposition and cementation, etc. *These are the processes that make up the rock cycle.* Conclude by outlining the rock cycle as a circular diagram of continuing processes.

Challenge the advanced groups to collect as complete a set of rocks representing the different varieties of each origin as they can. Refer them to available books to find out about the wide variety of rocks known to exist.

You may wish to have the advanced groups make rocks (by analogy). To make "sedimentary rock," add fine clay to a jar containing enough salt water to wet the clay thoroughly. Allow the clay to dry (heating will accelerate drying). The salt acts as a cement to bind the fine clay sediment into rock. "Igneous rock" can be made by melting moth flakes in a test tube and then allowing the melt to cool. Crystals of "igneous rock" will form. Changing the rate of cooling—by putting the melt in a refrigerator or by pouring it into cold water—will affect the size of the crystals which form. *Remember:* slow cooling forms big crystals, fast cooling results in small crystals. "Metamorphic rock" can be made by filling an envelope with moth flakes and squeezing it very tightly in the jaws of a vise. Leave it in the vise for at least 24 hours. Then examine its texture. Of course, each of these analogies has shortcomings and is meant only to demonstrate the kind of process involved.

6

Weathering and Erosion

MOTIVATORS

Why isn't the earth's surface entirely covered by solid rock? What happens to rocks exposed to the atmosphere which causes loose rock debris, dirt, and soil to form as a veneer over them? Without this cover of broken-up and altered rock material, vegetation would not be able to grow and without plants, animal life could not be sustained. Can you think of any animal which directly or indirectly does not rely on plant life for food?

BACKGROUND INFORMATION

Weathering, the mechanical and chemical breakdown of rocks, and *erosion,* the pickup and transport of weathering products, function "as a team" to alter the face of the earth's surface. *Mechanical weathering* includes processes such as rock cracking by the wedging action of freezing water, gradual heaving up of rock layers due to continued growth of plant roots into cracks, and rock fracturing both by natural processes (earthquakes, mountain building, etc.) and man's activities (quarrying, blasting, construction, etc.). *Chemical weathering* is largely a result of the interaction between water percolating through the ground and minerals in the rocks. Some minerals are dissolved by the water, others gradually decompose to form new minerals as they add water (hydration) or oxygen (oxidation), or alter chemically in other ways.

Weathering slowly changes solid rock into loose particles which can then be carried away by the *erosive action* of running water, wind, or glaciers. Some material is carried in solution in water, but most is transported in a form known as *sediment*. Sediment ranges from fine clay-sized particles held in suspension, to cobbles and even boulders which are tumbled along by strong currents.

No rock at or near the earth's surface can escape weathering and erosion to some degree. These rocks are continually undergoing change, and gradually the surface of rock is lowered as the upper layers of once-fresh rock are weathered to loose rock debris and soil and then gradually worn away by erosion.

MATERIAL

- Balances
- Beakers or widemouthed jars (9)
- Data sheet for recording measurements
- Graph paper
- Large nails (3)
- Pieces of limestone (3)
- Pieces of steel wool (3)
- Sea water or salt
- Vinegar or dilute hydrochloric acid

STUDENT PROCEDURE

The object of this experiment is to measure the changes in the weights of materials (limestone, nails, steel wool) subjected to different solutions (tap water, sea water, vinegar water or dilute hydrochloric acid) for a period of one or two weeks. Students can work in small groups, each with a different material-solution combination. In this way no two groups make identical studies, and a total of nine combinations can be observed and measured.

Prepare enough of each solution (sea water, vinegar water or dilute HCl) for use in the entire class. This is important so as to standardize each solution being used on the different materials. Use a mixture of 1 part vinegar to 4 parts tap water for the vinegar-water solution. If sea water is not readily available, prepare a sea-water solution using a ratio of 29 grams of sodium chloride (table salt) to each liter of water. Dilute HCl is prepared by a mixture of 1 part concentrated HCl to 9 parts tap water.

1. Carefully weigh the material (limestone, nail, or steel wool) given to you by the teacher. Record this as the "initial weight" on the data sheet (Table G-3).

Table G-3. *Record of weights of materials in different solutions.*

Material	Solution	Initial Weight	Second Weighing	Third Weighing, etc.	Final Weight
Nail	tap water				
Nail	sea water				
Nail	vinegar water (or dilute HCl)				
Limestone	tap water				
Limestone	sea water				
Limestone	vinegar water (or dilute HCl)				
Steel wool	tap water				
Steel wool	sea water				
Steel wool	vinegar water (or dilute HCl)				

2. Place the material in a beaker.

3. Pour the solution (tap water, sea water, vinegar water or dilute HCl) over the material so that the top of the material is at least 1 cm below the level of the solution. (This will allow the material to remain immersed, even with some evaporation during the span of the experiment.)

4. At prearranged times remove the material, allow to dry, weigh carefully, and return the material to the solution. (Your teacher will tell you how long the experiment will be conducted, how many times the material will be weighed, and at what time intervals. For best results, the experiment should be conducted for at least one week with no fewer than five weighings.)

5. Record the weight on the data sheet after each weighing.

6. After the final weighing, prepare a graph showing the weights recorded against time intervals.

ANALYZING RESULTS

Two of the controlling factors of chemical weathering can be studied by this experiment: (1) The changes observed in a nail, piece of limestone, or steel wool subjected to a solution over a period of time—thus *time* is the variable factor; (2) The changes measured and observed in like materials (nails, limestone, or steel wool) subjected to different solutions. The variable is the *environment* or chemical conditions as controlled by the nature of the solution—tap water, sea water, vinegar water or dilute HCl.

Natural conditions are, of course, considerably more complex. A wide variety of rock types exists with a range in susceptibility to weathering. And the chemical compositions of the solutions vary greatly, largely dependent upon the material with which they come in contact as they travel over and through the ground.

The significance of the quantitative changes should be discussed. The possible explanations for both increases and decreases in weight after weathering should be mentioned. Weathering may cause volume increases in rocks (for example, the addition of water to some minerals), although volume (and weight) losses such as may result from scaling off of weathered material or the removal in solution of some minerals are more characteristic.

Weight changes due to weathering are not necessarily uniform in rate. For example, limestone immersed in vinegar water or dilute HCl may show significant weight losses after the first one or two weighings only. Perhaps the acidity of the water has been greatly reduced after a few days, thus lowering its effect on the limestone.

The interdependence of weathering and erosion is quickly realized. Weathering makes the rock surface more vulnerable to erosion. Erosion removes the weathered material, thus continually exposing more rock to weathering processes. If erosion ceases, fresh rock is soon hidden deeply beneath a cover of its own weathered material and further weathering is impeded. If erosion is rapid, a fresh rock surface is constantly exposed.

INVESTIGATING FURTHER

Although erosion is a continual process, there are times when it is accelerated significantly. Find out how many of the students have witnessed a dust storm! To demonstrate this further, have the students observe a convenient creek or river during periods of normal flow and then during flooding or just after a heavy rain. Notice the difference in the color and velocity of flow of the water. Advanced groups can collect equal volumes of the water (in jars) during these contrasting conditions and allow the sediment in the water to settle to the bottom of the jars. Then have them dry and weigh the sediment that has collected. Why was there more sediment in the rain-swollen waters?

A field trip (either as a class or individually) to several stone buildings of different ages or to a local cemetery can greatly aid the understanding of rock weathering. Observe the surfaces of the tombstones (or stone buildings) and note differences. Do these differences seem to be a function of age or type of rock? Weathering has made it more difficult to read the inscriptions on the older tombstones. Stains, cracks, chips, and rough surfaces all increase with age of the exposure of these stones. Advanced classes should be challenged to identify the different types of rock they observe and to rank them on the basis of resistance to weathering with age. Beauty is only one factor for deciding on the kind of stone to use for tombstones and as veneer stone for houses. Another factor is the resistance to weathering—especially against staining caused by color changes during weathering of some of the minerals.

Students should not be left with the impression that weathering is necessarily a destructive process. True, quarrymen seeking fresh granite for building stone do not want weathered rock. But we are all dependent upon the products of weathering for much of our economy. Soil to grow our crops and clay minerals for a wide variety of industrial uses are just two examples. Have the class tabulate local industries using some products of weathering as useful materials to emphasize this fact.

43

7

Sediment Settling

With ever prevailing weathering and its inevitable consequence, erosion, why isn't the land surface soon entirely stripped of all sedimentary rocks? After erosion for hundreds of millions of years, would we not expect to see only those igneous and metamorphic rocks which originated at some depth below the earth's surface now exposed? Indeed, what happens to the products of weathering and erosion—minerals in solution, fine particles in suspension in running water, and the rock and mineral fragments which range from clay and silt to boulder size that can be seen in streams and rivers, along lake shores, and on the beaches of the oceans?

BACKGROUND INFORMATION

Loose, unconsolidated material that becomes the constituents of any sedimentary rock is known as *sediment*. Sediment, therefore, has a wide range in size and a great variety of compositions. It is the end product of the weathering process, and eventually all sediment ends up accumulated as deposits in some favorable setting such as in a lake or the oceans.

44

All sediment may be divided into two major types: (1) particles of pre-existing rocks and associated organic debris derived from weathering at the earth's surface, and (2) material obtained chemically and biochemically from precipitation out of water. Clay, silt, sand, and gravel are all examples of particles derived from previous rocks. Chemical precipitation from solution forms deposits such as limy muds (composed of calcium carbonate) and salt beds evaporated from saline-rich waters. Organisms utilize water to secrete shell and skeletal material (most commonly composed of calcium carbonate or silicon dioxide); after the organisms die, these shells and skeletons accumulate as biochemical deposits.

Sediment accumulates as the ability to transport it ceases. Most favorable as sites for deposition are the large bodies of relatively calm water, sometimes lakes but ultimately the oceans. Sediments tend to be deposited in nearly horizontal layers (the *Law of Original Horizontality*), generally deviating from this flat-lying accumulation only slightly. The tendency of sediments to be deposited in essentially horizontal layers results in the characteristic bedding noted in their ultimate product, sedimentary rocks. It is readily apparent then, that in a sequence of layers (either the original sediment layers or the beds of sedimentary rocks that result) the oldest deposits are at the bottom and each successively younger layer overlies the previous one. This concept is a fundamental one known as the *Law of Superposition*.

MATERIAL

- Clear glass jar with lid
- Sample of fine dirt (clay- to sand-sized particles)

STUDENT PROCEDURE

1. Place a handful of dirt into the jar and add water until about 4 cm from the top.
2. Firmly place the lid on the jar and shake the jar vigorously.
3. Allow the sediment to settle, observing the particles carefully as they accumulate. Is the sediment deposited in layers?
4. List all the features which can be observed from this experiment. Be sure to shake the jar several times to examine the results of the experiment repeatedly.

45

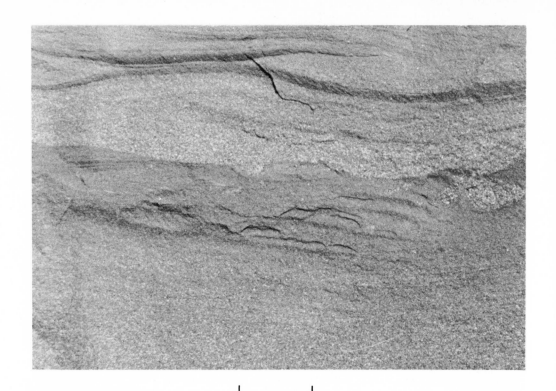

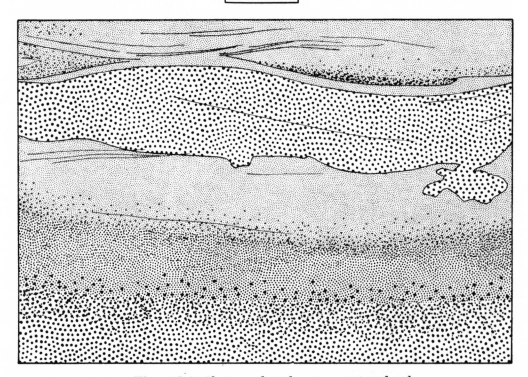

Figure G-6. *Photograph and accompanying sketch of graded bedding in sedimentary rock. Sample is right side up; grains grade from coarse to fine upward. Scale shown by bar which is 2 cm long. (Photograph courtesy of E. F. McBride.)*

ANALYZING RESULTS

Observant students will note a characteristic change in size of the sediments deposited. The largest particles will concentrate at the bottom and particle size will gradually decrease upward. This change in particle size from coarsest to finest can be observed within layers in sedimentary rocks and is known as *graded bedding*. Graded bedding results from the rate at which the particles settle. The largest particles generally settle fastest because they are commonly heavier than the finer particles (Figure G-6). If the finer particles were heaviest, they would be at the bottom and the change in graded bedding would be reversed.

Graded bedding is extremely helpful as supporting evidence that the Law of Superposition is valid. But suppose that the layers of sedimentary rock are tilted or even turned upside down (refer to Geology Experiment 9)? Graded bedding might prove that the sequence is overturned.

Properties other than particle size affect the layered appearance of the sediment in the jar. Most prominent is particle color. Another feature is the shape of the particles. All these features—size (and weight), color, shape—are related to the composition of the particles and, of course, ultimately to their source. Realizing this, is it then not possible to determine the probable source areas for the sediments which are being studied?

INVESTIGATING FURTHER

Have the students shake the jars and then allow settling with the jar tilted for a few seconds, then held upright. Some sediment will be deposited in a slightly tilted position. This is known as *cross bedding* and occurs when sediment is deposited on an angle, as on sand dunes and in some irregular stream channels (Figure G-7, next page).

To extend the experiment and make the analogy to deposition in nature more realistic, a larger "basin of deposition" like an aquarium can be utilized. Sediment mixed in a large container can then be added to the aquarium which should already have some water in it. Vary the kinds of sediment added by changing particle size, general color, etc. (This can usually be done easily by using dirt obtained from different places.) Different layers of sediment can then be observed with a graded bedding sequence (coarse to fine) for each added "deposit" of sediment.

To simulate cross bedding, prop up one end of the aquarium while

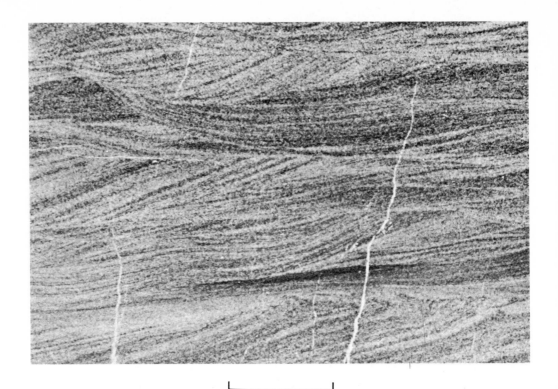

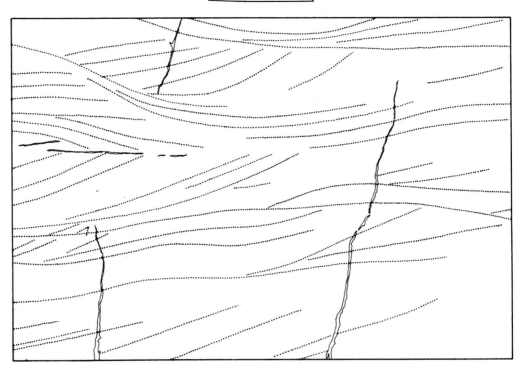

Figure G-7. *Photograph and accompanying sketch of cross bedding in sedimentary rock. Sample is right side up; cross beds flatten at the bottoms and are sharply truncated by overlying beds at the tops. Scale shown by bar which is 1 cm long. (Photograph courtesy of E. F. McBride.)*

you add one "deposit" of sediment. Then return it to a horizontal position before adding the next "deposit" which will bury the cross-bedded layer.

Challenge the most interested younger students and the advanced groups to find sedimentary rocks that reveal the same features observed in this investigation. Suggest that they find out where sedimentary rocks, especially sandstone, are exposed locally and look there.

8

Correlation of Geologic
Sections

MOTIVATORS

Layers of sediment harden into sedimentary rocks which can be traced by surface exposures across the land. This is a simple matter where individual beds of sedimentary rock are continually exposed. However vegetation and soil cover commonly obscure rock layers, and their continuity is interrupted by stream valleys and irregular topography. Under these circumstances, how can sedimentary rocks be correlated from place to place? And what does their distribution tell us about the basins of sediment deposition, location of ancient seas, and environments in the geologic past?

BACKGROUND INFORMATION

As sediments continue to accumulate in basins of deposition, they gradually attain thicknesses of 2000 to 8000 m and even more. Sediments in the lower layers, compressed by the weight of overlying material, are slowly converted to sedimentary rock. Compaction, dehydration, and some recrystallization occur as the sediment particles are firmly cemented together to form solid rock.

50

The types of sediment deposited, and thus the kinds of sedimentary rocks that form, are predominantly controlled by two things: (1) the kinds of rock from which the material comes, and (2) conditions such as depth of water, velocity of currents, etc. at the site of deposition. Coarser material (sand and silt) is commonly most abundant close to ocean shorelines and the resulting rocks are dominantly sandstone. Farther offshore, clay-sized particles may accumulate giving rise to shales. In quieter waters, limy muds commonly precipitate and from them limestones are formed. Thus a continual sequence of sedimentary rocks from limy muds some distance offshore to sand-sized particles along the beach may be deposited *at one time*. The locations where these different sediment types are deposited change as the shoreline shifts with advance or retreat of the sea. (For example, younger limy mud may accumulate over older sand or clay as the shoreline moves over the land.)

Gradually the basin fills with sediment and because of the shifts in the shoreline during deposition in the basin (or because of changes in the kinds of sediment being brought into the basin), the vertical sequence may consist of different kinds of sedimentary rock. Of course, the oldest rocks are on the bottom and the youngest on top, following the Law of Superposition.

The nature of the sedimentary rocks in a basin may thus give the geologist clues not only about the source rocks from which the sediments were derived, but also about the environment in the basin while the sediments were being deposited. Furthermore, changes in the rock types in a vertical sequence may help the geologist determine which way the shoreline was moving (landward advance or seaward retreat of the water). Of course, recognition of lateral as well as vertical changes in rock types is helpful in determining these shoreline shifts. Sequences also become an important factor in tracing layers of sedimentary rock from place to place for, whereas individual rock types may not be continually exposed, parts of the sequence may be recognized.

Figure G-8 illustrates the correlation of rock types, using information about their normal position in a basin to recognize the position of the ancient sea. It is true that the geologist generally observes many exposures scattered over a wide area, thus he may be able to outline the shoreward edge of the sea over a state or even recognize its entire extent across the continent.

MATERIAL

- Geologic sections (for example see Figure G-8, page 53).

51

STUDENT PROCEDURE

Give each student a copy of an illustration which contains three or four columns of rock layers or a continuous profile with spotted rock exposures (Figure G-8, Parts A and B or a comparable diagram).

1. Based on the relationships between types of sedimentary rocks and depositional environments (as explained by the teacher), draw lines correlating similar rock types laterally (as done in Figure G-8, Part C).
2. Indicate the directions of "shore" and "deeper water" and mark the shoreward edge of deposition of each rock type.
3. Outline the history of the area as revealed from the rock sequence and exposures.

ANALYZING RESULTS

Discuss the different interpretations presented by the students for the columns. By starting with a simple environmental condition, similar to the one in Figure G-8, the students have a limited number of possible interpretations. The exposures reveal a history of shoreline encroachment onto land. Then they proceed to a more complex series of columns which involve advance and retreat of the sea as documented by repetition of rock types. Introduce this by having the class sketch the rock types which would be found on the surface of Figure G-8, Part C if a record of shoreline retreat had been preserved after the shoreline advancement shown.

The kinds of fossils which occur in the rocks should also be considered as they are useful environmental indicators (refer to Geology Experiment 2). Thus animals living in shallow marine, shoreline, and beach environments (burrowing clams, oysters, crabs, etc.) commonly occur in sandstones. In contrast, limestone will contain evidences of those animals living in quieter, generally deeper water (echinoids, some corals, scallops, etc.). Realizing this, the students should be able to indicate those fossils common to each rock type from a list of fossils presented to them.

INVESTIGATING FURTHER

After the students have analyzed hypothetical examples, actual areas should be presented. The most desirable examples are *local* ones, based on exposures readily accessible to the school if possible. Advanced groups might be encouraged to examine these exposures on a field trip, or

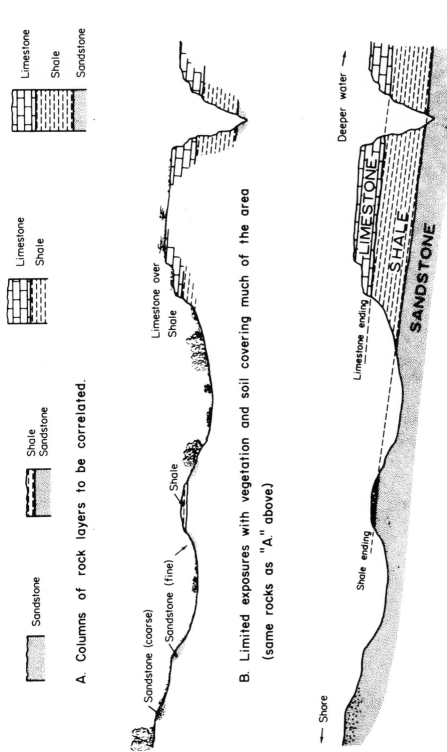

Sandstone

Shale
Sandstone

Limestone
Shale

Limestone
Shale
Sandstone

A. Columns of rock layers to be correlated.

Sandstone (coarse)

Sandstone (fine)

Shale

Shale

Limestone over
Shale

B. Limited exposures with vegetation and soil covering much of the area

(same rocks as "A." above)

← Shore

Shale ending

Limestone ending

LIMESTONE

SHALE

SANDSTONE

Deeper water →

C. Correlation of exposures based on sequences.

Figure G-8. *Correlation of columns of rock layers (A), or a continuous profile with spotty rock exposures (B), to form a picture of sea transgression (C), with sandstone nearest the shoreline and limestone seaward. Scale of horizontal distance shown may range from one to several kilometers.*

obtain the information from a geologic map of the vicinity or from other available local sources. They can then prepare actual profiles of the areas and sketch the rock types where found. Finally, environmental interpretations can be compared with fossils found locally to see if they agree. Disagreement might suggest an error in the interpretation. However it might indicate that the fossils were transported into this area after death along with the sediments, and that they did not live there. How else might you explain the presence of what seems like the "wrong" kind of fossils for a particular environment in rocks of that environment? (Explanations might include: incorrect identification of the fossils; adaptation of the animals or plants to a different environment through evolutionary changes; or a geologically "sudden" change in the environment, trapping the animals or plants which then die and are buried in sediments of the new environment.)

9

Folding and Faulting

W e have seen how rocks at the earth's surface are continually subjected to weathering and how the loose rock material, called sediment, is transported to favorable sites of deposition. Why, then, are new rocks continually being exposed and how do they get there? And why are some of these rock layers folded, bent, and broken—especially in mountainous regions?

BACKGROUND INFORMATION

After sediments are deposited, they are gradually buried by more sediments. As they are buried more deeply, pressures on them increase. These pressures cause the sediments to compact, forcing out water from between the particles, and as the particles are cemented together, make them into firm sedimentary rock.

Very high pressures exist in the deep interior of the earth—and accompanying these high pressures are high temperatures. Indeed, scientists believe that the outer core of the earth is liquid rather than solid. Between the earth's surface and the core, the rock is a solid although it is very hot and under high pressure and may deform by slowly flowing.[1] When this pressure is released, like during a volcanic eruption, the rock may melt and pour out on the surface as a lava flow.

Because of high pressures and temperatures and differential stresses within the earth, slow movements of some rock material are taking place below the surface. These movements may cause large masses of the crust to push slowly upward forming mountains. Layers of sedimentary rocks in these masses, with metamorphic and igneous rocks beneath them, are squeezed and contorted into *folds* and even broken and moved apart along *faults*. As rocks at the surface are stripped away by erosion, rocks once at progressively deeper levels in the crust are exposed.

Thus we see that the earth's surface is continually undergoing change —in some places slowly wearing down, in others gradually moving upward.

MATERIAL

- Modeling clay (three or four different colors)
- Plastic knife or thin wire
- Wood block (10 cm × 10 cm × 4 cm)
- Patterns accompanying this experiment (Figures G-9 and G-10)

STUDENT PROCEDURE

The various colors of clay represent layers of different sedimentary rocks.

1. Flatten the clay (by rolling out [2]) into layers approximately 1 cm thick. Cut out rectangles about 8 cm × 10 cm in size and place different-colored rectangles on top of each other to make "sandwiches" or blocks four or five layers thick.
2. Fold blocks into *anticlines* (convex upward bends) and *synclines* (concave upward bends) as shown in Figure G-11 (page 60) by compressing the sides.
3. Form a *dome* (an anticline with layers dipping away from a central high point) by slowly pushing a golf ball or other small, smooth object upward through the center of the clay block.

---

[1] An excellent film explaining this concept is ESCP's *How Solid Is Rock?* See Appendix V.

[2] A broom handle cut into 15-cm pieces makes excellent rolling pins.

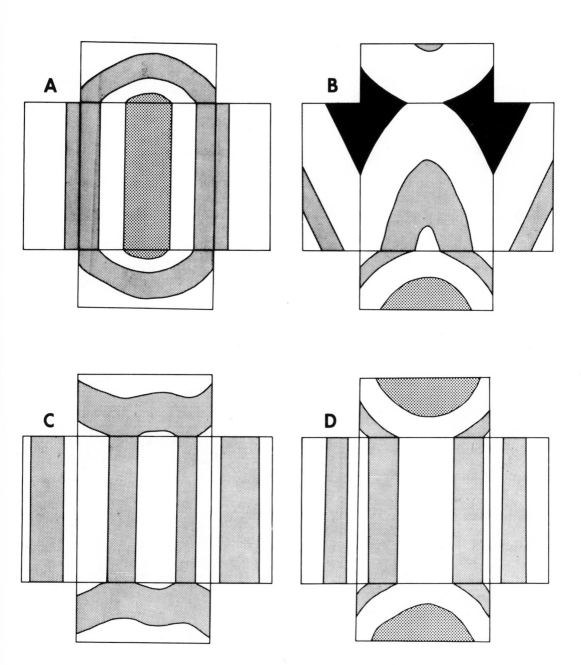

Figure G-9. *Patterns for block diagrams of fold types: (A) syncline, (B) plunging anticline, (C) an-ticline-syncline-anticline, and (D) anticline.*

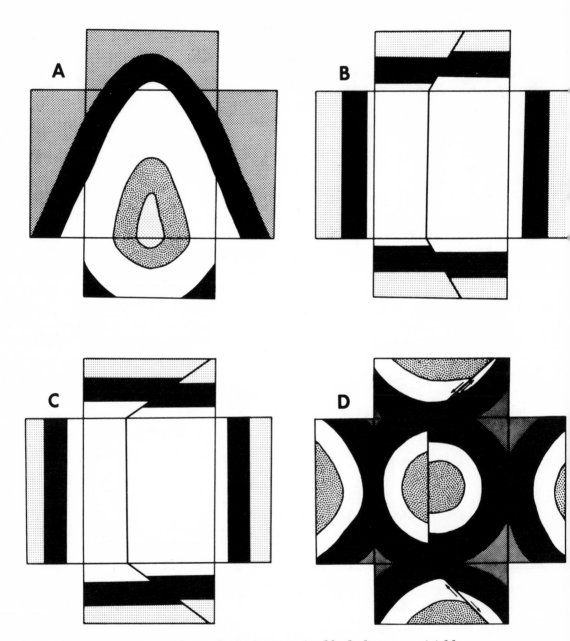

Figure G-10. *Patterns for block diagrams of fold and fault types: (A) plunging syncline, (B) normal fault of horizontal strata, (C) reverse fault of horizontal strata, and (D) dome cut by normal fault.*

4. Using a knife or piece of thin wire, cut through a clay block to represent a *fault*. Then slide the two pieces along the cut as during fault movement. Two of the most common types, a *normal* fault and a *reverse* fault, (shown in Figure G-11), can be reproduced in this manner.

5. Make more complex forms by first folding a clay block and then cutting it and moving the faulted blocks. For example, make a dome cut by a reverse fault or a syncline cut by a normal fault.

6. To show folds and faults by another method, wrap paper around a block of wood (styrofoam or other material can be substituted for the wood) and draw the views as would be seen for the top and side views. Begin this by using the patterns shown in Figures G-9 and G-10.

ANALYZING RESULTS

This experiment demonstrates several kinds of rock deformation that take place within the earth's crust. The activity also aids the students' ability to visualize in three dimensions so that they can "picture what is under the ground" by seeing what is on the surface or in road or stream cuts.

Insight should be gained into the *mechanisms* of rock deformation and this should be emphasized in classroom discussion. Anticlines and synclines generally form in response to compression (squeezing) of crustal rocks. In contrast, domes may be caused by material moving upward— resulting in extension (stretching) of the rock layers. One example is rising salt bodies which may produce domes in overlying sedimentary rocks.

Whether a fault is normal or reverse depends on the relative values of the stresses acting on the rocks to cause them to break. (Stress is simply defined as the *effects* of the forces acting on the rocks.) These stresses develop over a range of depths within the earth and are measured vertically and horizontally. Rocks break because they are too brittle to deform by bending or flow. Normal faults cause lengthening of rock bodies, whereas reverse faults result in their shortening. Geologists, therefore, realize that some parts of the crust undergo extension whereas other parts are compressed. Stresses in crustal rocks are perhaps best explained by suspected movements of material in the earth's lower crust and mantle.

INVESTIGATING FURTHER

The patterns included represent some of the basic forms of folds and faults. More complex forms can be devised by the students. For advanced

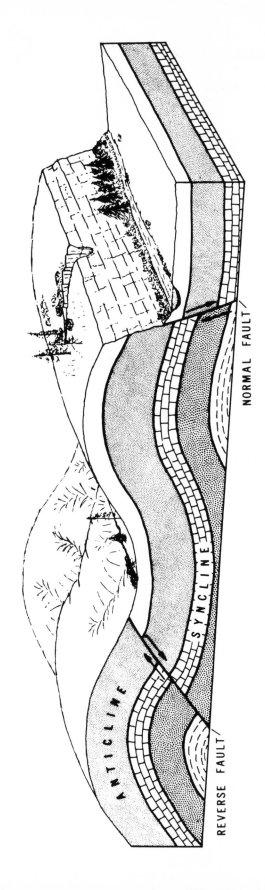

Figure G-11. *Block diagram of a segment of the earth's crust illustrating folds (anticline and syncline) and faults (normal and reverse). Note that a normal fault tends to lengthen the segment of the earth's crust shown, whereas a reverse fault shortens it.*

classes, patterns with one or more views missing can be utilized effectively. In order to visualize the missing views and draw them correctly, encourage the students to prepare clay blocks of these features. They can then copy the views seen on the patterns. An interesting challenge is to have each student prepare one complicated pattern with one or two views missing. Then have other students fill in the missing views by visualizing the feature.

It is important to emphasize that most of the fold and fault models the students design do indeed exist as real features in nature. Extremely complex forms can develop, especially in areas of mountain building.

Certain rules of age sequences become apparent after careful examination of the surface views of eroded folds. Oldest rocks occur in the centers of eroded anticlines with progressively younger rocks outward. However, the reverse is true for synclines as the youngest rocks are in the center. Do you see why? How might this be useful in recognizing the kinds of folds present when walking over them on the ground if side views are not available?

10

Interpreting Geologic History

MOTIVATORS

The Laws of Original Horizontality and Superposition (Geology Experiment 7) outline the nature of deposition for sedimentary rocks. These rocks may then be deformed (folded and faulted) as described in the previous experiment. What other things can happen to rocks during their long history of existence? How might the complete sequence of events that took place in an area be interpreted?

BACKGROUND INFORMATION

Upward movement of plutonic igneous rock is commonly associated with the folding and faulting of overlying sedimentary rocks. Igneous bodies may move upward as huge masses known as *stocks* and *batholiths*, lifting up the overlying sedimentary rocks. Or the magma may invade these layers, flowing along bedding planes, faults, and other planar features to form long tabular bodies called *dikes* and *sills*. (Dikes cut across adjacent bedding in the sedimentary rocks, whereas sills parallel this bedding.) Attendant with magma intrusion is metamorphism of the invaded rocks, causing metamorphic rocks to form, generally in zones adjacent to the igneous bodies.

Subsequent erosion may remove the uppermost layers. Erosion may continue indefinitely, exposing the rocks at continually greater depth, or the region may subside and once again be under water and thus covered by sedimentary deposits. If so, the new influx of sediments will bury the previous land surface. However this buried surface, known as an *uncon- formity,* remains as evidence of its former existence above the seas.

By viewing sections of the earth's crust, the order of events, and thus the geologic history of the area, can be established. The major events to observe are deposition, folding, faulting, and erosion with uplift accompanying erosion and deposition following subsidence.

MATERIAL

• View of a section of the earth's crust (e.g. Figures G-12 and G-13).[1]

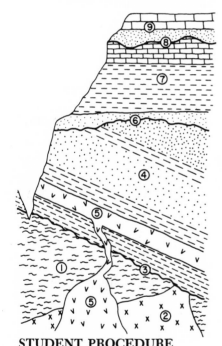

Figure G-12. *Section view of lower rocks of the Grand Canyon, Arizona. Major geologic events in sequence are: (1) Deposition of sediments; (2) Intrusion of igneous rock; sedimentary rocks folded; metamorphism of sedimentary rocks; (3) Uplift and erosion; (4) Subsidence and deposition of sediments; (5) Intrusion of igneous rock bodies; (6) Uplifting and tilting of sequence; erosion; (7) Subsidence and deposition of sediments; (8) Uplift and erosion; (9) Subsidence and deposition of sediments. Vertical distance shown is over 2 km.*

STUDENT PROCEDURE

1. Study the section view of the Grand Canyon area shown in Figure G-12.

---

[1] The Geologic Map Portfolios published by Williams and Heintz Map Corporation afford excellent resource material for maps and section views of segments of the earth's crust. See Appendix V.

2. List the events documented by this section view in order of their occurrence. Be sure to include uplift with erosion and subsidence with deposition to convey relative height with respect to movements of the sea.

3. Do the same thing with the section view along the flank of the southern Rocky Mountains, Colorado (Figure G-13).

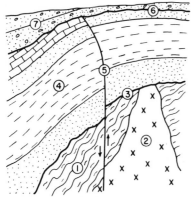

Figure G-13. *Section view of rocks along flank of southern Rocky Mountains, Colorado. Major geologic events in sequence are: (1) Deposition of sediments; (2) Intrusion of igneous rock; sedimentary rocks folded; metamorphism of sedimentary rocks; (3) Uplift and erosion; (4) Subsidence and deposition of sediments; (5) Some folding, fault develops; (6) Uplift and erosion; (7) Subsidence and deposition of gravels; and finally uplift and erosion. Vertical distance shown is approximately 2 km.*

ANALYZING RESULTS

The major geologic events are listed for both the lower Grand Canyon section and the southern Rocky Mountain section (Figures G-12 and G-13 respectively). Erosional surfaces such as numbers 3, 6, and 8 of the Grand Canyon section can be readily recognized by the irregularity of the surface, reflecting the nature of the topography before subsidence and burial, and by the contrast in rock types above and below this surface.

It is common to find sandstone (represented by a dot pattern) as the first deposit after sea advance—note the position of sandstone layers in the sedimentary sequences in Figures G-12 and G-13. Sandstone commonly grades upward into shale (pattern of short dashes) which can again be recognized in these examples. (Recall the discussion of environmental control on types of sediment deposited relative to shoreline positions in Geology Experiment 8.)

The time of the last fault movement can readily be documented because a fault plane cuts and displaces those rocks that it antedates; rocks deposited or intruded after faulting are not cut by the fault plane. However earlier movements along a fault plane may be difficult or impossible to distinguish.

By observation of the same event (such as intrusion of igneous bodies, faulting, or uplift) from several section views of the rocks, its areal extent can be approximated. Events which were significant over wide geographic areas should be evident from exposures over extensive areas, whereas local events will be restricted to a few closely associated exposures. By this technique, the region involved in any one event can be accurately delineated. Thus the areal distribution of a sea during a particular geologic time or the area involved in an episode of mountain building can be established.

INVESTIGATING FURTHER

Have each member of the class prepare a hypothetical section view of a segment of the earth's crust showing a sequence of three or four geologic events and have the rest of the class interpret these views. Classes can be presented with the sequence of events in the geologic history of an area (more complicated histories for more advanced groups), and asked to prepare schematic views of these events. If the information is available, the geologic history of the local area should be studied in this manner. Where possible, the interpretation of local geologic history should be "field checked" by examining rock exposures that can be visited. In such cases, it is unlikely that any one exposure will reveal more than a segment of the history; hence, generally several areas must be visited and the history "pieced together" from these areas.

From this discussion do you see why "classic exposures" like the deep river canyon cuts (Grand Canyon, Arizona; Black Canyon of the Gunnison River, Colorado; Yellowstone River Canyon, Wyoming) or the large areas of well-exposed rocks (Bryce Canyon National Park, Utah; Death Valley, California) are so important to geologists in their study of earth history? How might the study of rock cores obtained in drilling aid geologists' understanding of the geologic history of the areas drilled?

11

Making a Topographic Map

MOTIVATORS

Maps are made for many purposes, for example for locating things, identifying many different objects, and seeing features in relationship to their surroundings. Topographic maps are useful for all these purposes for viewing features on the earth's surface—both the natural conditions and the modifications and additions imposed by man. Do topographic maps offer any other advantages to the viewer? Are these maps only useful to geologists or may they be of value to others—hikers, hunters, farmers, real estate developers, tourists?

BACKGROUND INFORMATION

The configuration of the land surface is known as the *topography*. The method most commonly used to represent topography on a map is by drawing lines connecting points of equal elevation. Called *contour lines,* they are numbered to indicate the true elevation of the points they connect. Because sea level is a common elevation available on all continents, it is generally used as the zero contour line. Numbering is positive above sea level and negative below.

The number spacing between contour lines is known as the *contour interval*. In areas of great relief (high peaks and deep valleys) the contour interval may be 100 ft or more. For areas of low to moderate relief, an interval of 20 ft or even 10 ft is common. (Note common use of English system on U. S. maps.)

The following "rules" apply to contour lines on topographic maps and should be considered:

(1) The closer together the contour lines (of a given contour interval), the steeper the slope represented.

(2) Contour lines never cross or intersect. (*Exception:* a vertical or overhanging cliff.)

(3) Contour lines are parallel (or nearly parallel) where the land slope is constant.

(4) Contour lines bend forming a "V" which points uphill where they cross narrow low areas, for example stream valleys.

(5) Contour lines are always numbered at a uniform interval so that they can be numbered back to zero (sea level).

(6) Closed contour lines (such as circles and ellipses) represent high areas—hills, peaks, etc. The exception is closed contour lines of depressions which are always hachured to distinguish them. (A hachure is a short line extending from the contour line toward the lower elevation, ⊤⊤⊤⊤⊤.)

MATERIAL

- Relief model of landform to be contoured [1]
- Plastic shoebox, including lid
- Grease pencil
- Food coloring
- Topographic map [2]

STUDENT PROCEDURE

1. Place uniformly spaced marks (vertically) on one side of the shoebox from bottom to top. The spacing between marks represents the contour interval.

---

[1] Hubbard Scientific Company sells an excellent model designed for this purpose. See Appendix V.

[2] Refer to U.S. Geological Survey series of topographic maps. Preferably use a map of local area if available. See Appendix V.

Figure G-14. *Topographic map taken from the Clyde Quadrangle, New York (15-minute series) located just south of Lake Ontario. The elongate hills are drumlins (mounds of rock debris) deposited as a glacier of Pleistocene age moved southward (evidenced by the direction of drumlin elongation) over the region. Bar for scale is 1 km; map oriented with north at top (Map from U. S. Geological Survey.)*

Figure G-15. *Topographic map taken from the Milton Quadrangle, Pennsylvania (15-minute series) located in the northeast central part of the state. The east-west trending ridges are part of the Ridge and Valley Province (a result of anticlinal and synclinal folding) of the Appalachian Mountains. Bar for scale is 1 km; map oriented with north at top. (Map from U. S. Geological Survey.)*

2. Place the model in the box and add dyed water up to the first mark from the bottom.

3. Place the lid on the box and *viewing from directly above,* draw a line (representing a contour line) on the lid with the grease pencil following along the water-model contact.

4. Then add more water to the next highest mark and draw another line on the lid.

5. Repeat this procedure with each interval marked on the side of the box until the model is covered with water.

6. Assign an interval to the spacing between contour lines and number the lines beginning with zero (sea level) for the lowest line.

7. Then examine the topographic map to identify the features on the model as they are shown on the map.

ANALYZING RESULTS

Based on their observation of the topographic maps just made, have the students list all the features they can about contour lines and topographic maps. Most of the "rules" described previously (in Background Information) should be listed independently by the students. Discuss the list of student observations and each "rule" about contour lines.

Then raise the question, what features are unique to topographic maps that are of special value to geologists? The idea that topographic maps present three-dimensional portrayals of the land surface should be emphasized. Realizing this, the question of major importance then becomes, "What geologic phenomena cause the present topography?" Of course, an interplay of several factors is at work over any one area to cause the features observed. These include climate, nature of bedrock, structural features present (folds and faults), amount of vegetation, length of exposure of this area since uplift, extent of stream-pattern development, etc. In addition, special conditions may exist or have existed which contribute to the topography. Extrusion of volcanic material (volcanic cones produce obviously noticeable features), erosional or depositional products of glaciers having passed over the area, collapse of segments of the land in areas of limestone bedrock resulting from cavernous solution activity, and the piling up of wind-blown debris as sand dunes are all examples of these special conditions.

Conclude the discussion by mentioning the many nongeological uses to which topographic maps might be applied.

INVESTIGATING FURTHER

Obtain and study topographic map(s), with the following questions in mind:

(1) What is the significance of each of the colors on the map? (Standard colors used are: black for man-made or cultural features; blue for water or hydrographic features; brown for relief; green for woodland cover; and red emphasizes important roads, shows built-up urban areas, and public-land subdivision lines.)

(2) What is the scale of the map? Why is it represented in different ways? (It is standard to have the scale in both English and metric systems for use by persons conditioned to either system. The scale is also commonly expressed as a fraction.)

(3) What is the contour interval?

(4) Locate the points of highest and lowest elevation on the map. How is each formed?

(5) What are the latitude and longitude boundaries of the map?

(6) Locate as many different landforms as you can on the map. (For example: lake, level land, peak, ridge, steep cliff, valley.)

(7) What factors or special conditions may exist (or existed in the past) to control the landforms prominently displayed?

Then have the students draw a profile between two points assigned on the topographic map. The procedure for drawing profiles is explained in several of the secondary school earth science textbooks. For an especially thorough discussion of topographic maps, refer to ESCP Reference Series Pamphlet No. 7 and the literature on topographic maps available from the U. S. Geological Survey, listed in Appendix V.

12

Geology from Aerial Photographs

MOTIVATORS

Topographic maps enable three-dimensional visualization of the configuration of the land as illustrated in the previous experiment. However, photographs of the ground surface taken from airplanes may serve this purpose even better. When two photographs of the same area (taken from different camera positions) are viewed simultaneously, the three-dimensional shape of the land can be observed. Can you see why geologists tracing faults, folds, and rock types commonly rely on aerial photographs as an important aid to mapping? How might aerial photographs taken daily be of importance in a military effort?

BACKGROUND INFORMATION

Aerial photographs can be generally divided into two main types, oblique and vertical. *Oblique photographs* are taken with the camera system directed toward the ground surface at an angle from vertical. *Vertical photographs* are taken with the camera system directed essentially vertically downward. Vertical photographs are most useful to geologists because they afford three-dimensional viewing (oblique photographs can only be viewed three-dimensionally with great difficulty), and have the perspective most useful to mapping the rocks—the same view as shown by a geologic map.

To view photographs in three dimensions, known as *stereoscopic viewing,* overlapping pairs of photographs are required. (Generally the overlap of the photographs is approximately 60%.) Only the overlapping

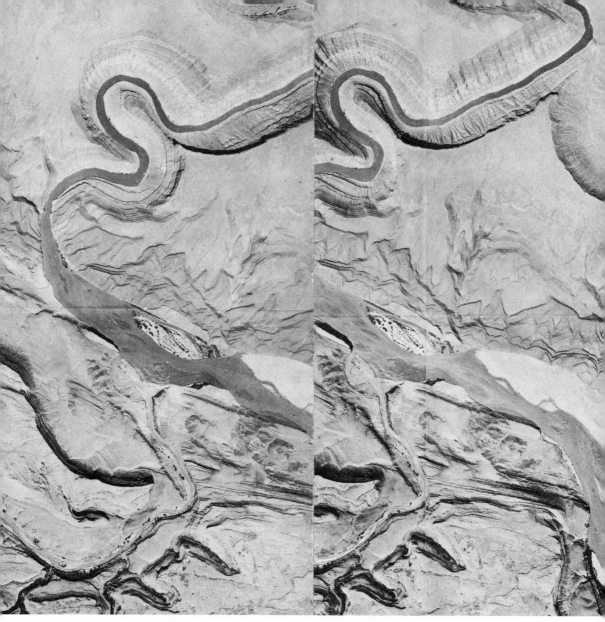

Figure G-16. *Aerial stereo-photograph of an area in San Juan County, Utah. Horizontal sedimentary rock layers are exposed where the San Juan River flows in a deep gorge at the top of the picture. (Note the banded pattern on the valley walls caused by the horizontal bedding.) The deeply incised meandering of the river valley is a result of downcutting as the land was rising. The layers bend sharply into a nearly vertical position near the bottom of the picture. The steeply dipping layers can be identified by the sharp ridges and the straight-line pattern of the rocks. The fold formed as the horizontal layers (top of picture) bend into nearly vertical layers (bottom of picture) is called a* monocline. *Approximate scale of photograph 1 cm = 200 m. (Photograph obtained from Cartographic Branch, National Archives, Washington, D.C.)*

portion (the part shown on both photographs) can be viewed stereoscopically. By placing the two photographs close together, the same features on each can be observed simultaneously, one eye on each photograph. Normally, one's line of sight converges when objects are viewed at close range. However to see stereoscopically, viewing must be along essentially parallel lines of sight. A *stereoscope* is commonly used to aid this technique of viewing. (The simplest and least expensive stereoscopes are constructed with two small lenses in a frame. The stereoscope is placed directly on the photographs and viewing can be done comfortably from a seated position.)

Geologic interpretation of aerial photographs involves careful observation of the features, including their color tones, distribution and shapes, and then relating them to known geologic features. Experience in studying aerial photographs and some basic geologic knowledge are essential to acquiring the full benefit of information that photographs contain. Nevertheless, students can learn much about an area by studying aerial photographs, even though they have very limited backgrounds in geology.

MATERIAL

- Aerial photograph pairs [1]
- Stereoscope [1]

STUDENT PROCEDURE

Consider each of the following basic elements of the features observed on the photographs. Based on your observations, try to identify the features and tell from what and how they are made.

(1) **Size**—The height as well as the horizontal extent of the feature should be considered in making an identity.

(2) **Shape**—The general form commonly reveals the feature's identity. Shape is especially useful for man-made features and well-defined landforms.

(3) **Shade or Tone**—The tone of the object is usually a shade of gray. It results from the amount of light that the object reflects.

---

[1] Hubbard Scientific Company sells a spiral-bound manual entitled *Aerial Stereo Photographs* and an inexpensive pocket stereoscope. Other sources of photographs and stereoscopes are also available. See Appendix V.

(4) **Texture**—The general appearance (speckled, mottled, rough-looking, smooth, etc.) may indicate something about the makeup of the feature.

(5) **Pattern**—The arrangement of the features may be significant, especially if the pattern is strongly oriented, such as a parallel alignment of ridges or a series of hills with the steep sides all facing in the same direction.

(6) **Unique Characteristics**—Anything different which you notice about the feature should be considered of possible importance. The contrast with other features commonly reflects a different makeup or origin.

ANALYZING RESULTS

Begin by finding the location of the aerial photographs being studied. Many times, knowing where the photographs were taken is an important clue to the kinds of features to be seen. (For example, sand dunes in desert areas, glacial features in northern U. S. and Canada, volcanoes along the northwest Pacific states, folded and faulted rocks in mountainous regions, etc.)

Then discuss each of the basic elements previously listed. Remember that tone may be a response to the bedrock *or* the vegetation which covers it. Rocks covered with trees have a different shade than those with grass or barren rocks. Indeed, the different kinds of trees commonly vary in tone because of the nature and density of their foliage. Patterns might reflect trends of features caused by directional controls—prevailing winds aligning sand dunes, the direction of ice flow forming oriented depositional features, the layers of bedrock controlling the directions which the streams flow, etc. Keep in mind the normal sizes of features you can identify (buildings, airstrips, towns, roads, etc.) and compare them with the features you are attempting to identify.

The following geologic features should be considered for discussion if the appropriate aerial photographs are available: flat-lying beds, tilted beds, faults, folds, joint patterns, kinds of rock, volcanic features, wind deposits, glacial features, stream patterns, lakes, coastline features, cycles of stream erosion, and the uses of the land by man.

INVESTIGATING FURTHER

Have each student in the class select a pair of aerial photographs that illustrates a particular geologic feature. (*Aerial Stereo Photographs* lists the most prominent features shown on each plate in the manual.) Then have them prepare a series of questions about the feature which they have chosen.

An example, taken from Plate 61 in *Aerial Stereo Photographs,* deals with layers of different rock types. Questions are as follows:

(1) What controls the tones of the different bands of rock? *Answer*: the different rock types *and* the amount and kinds of vegetation on the different rock types. Note that some rock layers support abundant vegetation, others sustain very little.

(2) What use is made of this land? *Answer*: Farming in the lower flat areas (note the contouring of the land being farmed); ranching in the hill country.

(3) What evidence can you cite that there are no folds or faults in these rocks? *Answer*: The rock layers can be traced continuously across the photograph without break and without changes in elevation.

(4) Can you locate the area of youngest rock on the photograph? *Answer*: 1.7-B.7. The Law of Superposition (see Geology Experiment 7) indicates that youngest rocks are on the top of a sequence and the beds at this location are on top.

(5) How would you distinguish the limestone layers from the "bands" of shale on the photograph? *Answer*: The shale has a darker tone because the rock and its soil are darker than limestone.

After the students have prepared their questions, a class discussion can be held to answer them.

Advanced groups should be encouraged to use topographic maps in conjunction with aerial photographs of the same areas (if available) or of areas showing similar features. Comparison can then be made between the features as observed on the photographs and as they are shown by contour lines on the topographic maps.

Section Two

OCEANOGRAPHY

1

Mapping the Ocean Floor

MOTIVATORS

More than 70% of the land surface of the earth lies buried beneath the oceans. Although man has traveled on the oceans for centuries, only within the past few decades— especially the last 15 years—have scientists concentrated on obtaining knowledge about the mysteries the oceans contain. Oceanographers (scientists who apply their knowledge to studying the oceans) now realize that huge mountains exist on what was once thought to be a nearly flat ocean floor! But how do oceanographers obtain information about the deep-ocean floor? It is too deep for submarines; they generally travel at depths less than 200 m below the surface. What importance will the new findings of oceanographers have on our better understanding of the origins of the continents?

BACKGROUND INFORMATION

One way to measure ocean depth is to put a weight on the end of a string and lower the weight into the water. When the weight touches the ocean floor, the length of the string used is the water depth. But can you imagine using this method to measure the Nero deep in the

Mariana Trench of the Pacific Ocean—the oceans' deepest measured spot at 11,035 m? Scientists use a technique called *echo sounding* which records the time for a sound signal to travel from a ship to the ocean floor and back to the ship.

Knowing that sound waves move through sea water at an average speed of about 1460 m/sec, it is easy to compute total depth when the time of the echo sounding is recorded by using the formula:

$$\frac{Time \times Speed}{2} = Depth \ .$$

Ships generally don't make single sound readings. Rather, the ship makes a traverse from one place to another, continually sending out sound waves and recording their return as the ship moves on course. In this way, oceanographers obtain many depth readings in a line from which a *profile of the topography* of the ocean floor can be drawn.

MATERIAL

- Graph paper
- Data sheet of echo soundings (Table O-1)

STUDENT PROCEDURE

Refer to the data sheet which gives: (a) the horizontal distance of each sounding in kilometers from the point where the traverse begins, and (b) the echo-sounding time in seconds.

1. Begin by computing the depth of the ocean floor (in meters) for each of the 25 recorded soundings. For the first recorded sounding:

$$\frac{5.2 \ sec \times 1460 \ m/sec}{2} = Depth$$

$$3796 \ m = Depth \ .$$

2. List the depths in column four of the data sheet.
3. Plot the depth of each recorded sounding on the graph paper (locating points approximately). Use a horizontal scale of 1 in (on the graph paper) representing 10 km and a vertical scale of 1 in equals 2000 m. Be sure to plot the vertical scale with zero (sea level) at the top.
4. Connect the plotted points of each of the 25 recorded soundings.

The line connecting these points is the profile of the ocean-floor topography along the line of the traverse.

Table O-1. *Data for determining the ocean depth for a series of 25 recorded soundings of a hypothetical profile of the ocean floor.*

Recorded Sounding	Horizontal Distance from Origin (in km)	Echo-Sounding Time (in sec)	Depth to Ocean Floor (in m)
1	4	5.2	3796
2	8	5.4	
3	12	5.6	
4	16	5.8	
5	20	5.8	
6	24	4.0	
7	28	5.9	
8	32	3.0	
9	36	3.6	
10	40	5.8	—
11	44	5.9	
12	48	6.0	
13	52	6.2	
14	56	6.2	
15	60	4.2	
16	64	5.8	
17	68	6.0	
18	72	6.2	
19	76	2.6	
20	80	5.4	
21	84	2.8	
22	88	4.2	
23	92	3.0	
24	96	5.4	
25	100	5.2	

ANALYZING RESULTS

Only recently have oceanographers recognized what is shown by the profile you have just made—that the topography of the ocean floor is highly irregular with many peaks, ridges, troughs, and valleys.

Raise the question of the significance of the spacing between recorded soundings to the accuracy and detail of the profile. The profile just made extends over a horizontal traverse distance of 100 km and included 25 recorded soundings, one reading each 4 km. Suppose soundings were recorded at 2-km intervals, then at 1-km intervals. Control of the topography and accuracy of the profile would steadily improve as the sounding interval decreased. Can you see why oceanographers want the ship to move slowly and have an essentially continuous profile recorded?

INVESTIGATING FURTHER

Discuss the similarities and differences that might be expected to exist between the topography on the continents and on the ocean floors. What agents of erosion are actively modifying the land areas of the earth's surface? Are these same processes at work modifying the ocean floors? (Oceanographers do indeed make analogies between erosion on land and that under the oceans. Currents are able to cause erosion, especially when they move rapidly and are heavily laden with sediments. Of course, features like wind and glaciers which can be observed as erosional agents on land are not present on the ocean floors.)

What possible value is the knowledge of ocean-floor topography to oceanographers? After all, we probably won't be laying pipelines on the ocean floor nor will we be constructing roads over this topography. To realize the value of this information, perhaps we need to examine a more accurate profile of an ocean floor. In the next experiment we study the profile of a traverse across the Atlantic Ocean.

2

Profile Across the
Atlantic Ocean

Ask the students what they think the ocean floor would look like if they could observe it without its water cover. What would the scenery be like driving across the floor of the Atlantic Ocean? Would they see rugged, towering mountains, steep-sided cliffs, and deep canyons, or would it be a flat plain?

BACKGROUND INFORMATION

The Atlantic Ocean floor consists of three major morphological divisions: (1) continental margin, (2) deep-ocean floor, and (3) mid-oceanic ridge.

The continental margin includes three provinces: (1) the submerged continental platform called the continental shelf, (2) the steep edge of the continental block called the continental slope, and (3) the raised edge of the ocean floor known as the continental rise.

Averaging less than 200 m, the *continental shelf* is a shoal (shallow underwater surface) of low local relief which extends from shoreline to the shelf break where the seaward slope of the ocean floor increases sharply. The shelf break or shelf edge delimits the continental shelf.

83

The origin of continental shelves is hypothesized to represent a combination of contemporaneous wave action and river erosion occurring when sea level was lowered during different stages of the Pleistocene Ice Age.

Exploration by petroleum geologists has revealed the existence of large accumulations of oil and natural gas contained within sediments on the continental shelves. The extensive development of oil and gas off the California, Louisiana, and Texas coasts has added greatly to U. S. fuel reserves. The shelves are also considered as future sources of other natural resources, such as sulfur, phosphate, manganese, and iron ore. Thus the continental shelves are rapidly becoming an important economic asset to the nations they border.

Continental slopes are the most imposing features on the earth's surface; they are the longest and highest continuous boundary walls in the world. The *continental slope* is a relatively steep but still gently dipping (3° to 6°) portion of the sea floor which extends from the shelf break at the outer margin of the continental shelf, to a gentle slope that extends for hundreds of kilometers into the deep-ocean basins.

The main topographic features of the continental slopes are submarine canyons which dissect it deeply in many localities, and commonly encroach onto the continental shelf. Geologists believe these canyons were formed by erosive undersea avalanches called turbidity currents. The continental slopes are formed principally by diastrophic activity (earth movements) occurring at the junction between the continental block and the deep-ocean floor.

The broad zone of gentle slope between the continental slope and the deep-ocean floor is called the *continental rise*. The seaward termination of the continental rise is generally abrupt, and at this boundary regional slopes decrease considerably.

The deep-ocean floor, the second basic morphological subdivision, includes abyssal plains, abyssal hills, guyots, seamounts, seamount groups, and mid-ocean canyons as important topographic features. It lies between the continental margin and the mid-oceanic ridge.

The *abyssal plains* are flat surfaces on the deep-ocean floor built primarily by turbidity current deposits. The *abyssal hills*, relatively small topographic features with relief usually less than 900 m, represent the unburied portions of the deep-ocean floor.

Other major relief features are:

Mid-Ocean Canyons—Steep-walled, flat-floored depressions occurring on the abyssal plains.

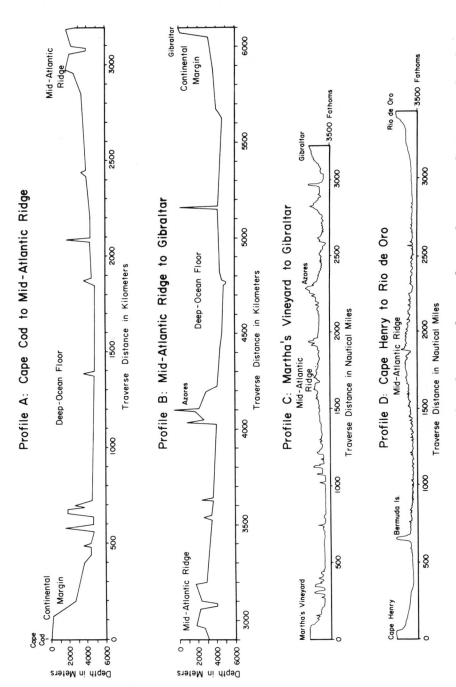

Figure O-1 *Topographic profiles across the Atlantic Ocean. Profiles A and B, drawn from the data listed in Table O-2, show the generalized configuration of the Atlantic ocean floor and illustrate the major morphological divisions. Profiles C and D were drawn from soundings continuously recorded by an NMC echo sounder on the R. V. Atlantis (reproduced from "The Floors of the Ocean: I. The North Atlantic," by Bruce C. Heezen, Marie Tharp, and Maurice Ewing, Geological Society of America Special Paper 65, plate 22).*

Seamounts—Submarine mountains generally considered to be of volcanic origin with comparatively steep slopes, relatively small summit areas, and relief greater than 900 m.

Seamount Group—An elongated series of seamounts.

Guyots—Flat-topped seamounts, also called tablemounts.

The *mid-oceanic ridges* are continuous median ridges running the length of the North Atlantic, South Atlantic, Indian, and South Pacific Oceans, for more than 64,000 km covering an area almost equal to the surface area of the continents.

That portion of the mid-oceanic ridge which lies within the limits of the North and South Atlantic Ocean is called the *Mid-Atlantic Ridge*. It consists of a broad fractured arch which occupies approximately the central half of the ocean.

The Mid-Atlantic Ridge is subdivided into two types of provinces: (1) the crest provinces, consisting of the rift valley and rift mountains, and (2) the flank provinces, located between the crest provinces and the deep-ocean floor.

The most striking feature of the Mid-Atlantic Ridge is the *rift valley* (Mid-Atlantic Rift), or long axial cleft in the crest of the ridge. The floor of the Rift usually lies at a depth of about 3500 m and has a width of about 20 km. Mountains several hundred meters high protrude from its floor where the valley is widest.

Steep walls flank the rift valley and form large rough-sided blocks known as the *rift mountains*. The rift mountain topography is extremely rough with steep, narrow, intermontane valleys and a local relief of about 700 m from valley floor to adjacent peak.

The *flank province* is characterized by rough topography with peaks more than 350 m high occurring at a frequency of about 10 per 100 km. Many of the valleys between peaks are smooth and slope away from the crest of the ridge.

Two trans-Atlantic profiles are reproduced in Figure O-1, page 85. These profiles illustrate the major morphological divisions and their provinces. Analysis of the profiles illustrates the diversity of expression and arrangement of the topographic features of the ocean floor. These profiles were originally plotted from soundings using a vertical scale of 2 in equals 1000 fathoms and a horizontal scale of 2 in equals 40 nautical mi, which resulted in a 40:1 vertical exaggeration. (One fathom = 6 ft or 1.8 m and 1 nautical mile = 6076 ft or 1852 m.) The original profiles were then reduced photographically.

MATERIAL

- Graph paper
- Data from Table O-2—depth and distance of traverse

86

Table O-2. *Data for determining a profile of the floor of the Atlantic Ocean. Profile A extends from Cape Cod to the mid-Atlantic; profile B extends from the mid-Atlantic to Gibraltar.*

POINT	DEPTH (M)	DISTANCE (KM)	POINT	DEPTH (M)	DISTANCE (KM)
\multicolumn 6 PROFILE A: CAPE COD TO MID-ATLANTIC			PROFILE B: MID-ATLANTIC TO GIBRALTAR		

1	0	0	*1	3100	2900
2	200	120	2	2700	2960
3	2700	200	3	1800	2975
4	3700	400	4	2200	3060
5	4400	440	5	4000	3075
6	4400	480	6	4000	3090
7	3700	490	7	2400	3100
8	4400	500	8	1800	3190
9	4600	560	*9	2900	3200
10	1600	580	10	3500	3525
11	4600	600	11	2700	3540
12	4600	640	12	3500	3550
13	1800	660	13	3700	3625
14	1800	680	14	2400	3630
15	3700	690	15	3700	3640
16	2700	700	16	4000	4025
17	4600	725	17	900	4035
18	4800	1375	18	2700	4050
19	3700	1390	19	1800	4090
20	4800	1400	20	+500 **	4100
21	4900	1850	21	2200	4110
22	3700	1875	22	2700	4190
23	4600	1890	23	4100	4225
24	4400	2075	24	4400	4425
25	1800	2090	25	4600	4500
26	4400	2100	26	4800	4750
27	4400	2210	27	5100	4760
28	4000	2420	28	5100	4770
29	3500	2440	29	4600	4780
30	4000	2450	30	4400	4825
31	3500	2850	31	4200	5150
*32	3100	2900	32	200	5160
33	2700	2960	33	4200	5170
34	1800	2975	34	4400	5390
35	2200	3060	35	4600	5625
36	4000	3075	36	4000	5675
37	4000	3090	37	3700	5875
38	2400	3100	38	2700	6050
39	1800	3190	39	200	6075
*40	2900	3200	40	0	6100

*Points 32 to 40 of Profile A, duplicate points 1 to 9 of Profile B.
**Above sea level.

STUDENT PROCEDURE

This experiment can be done effectively by working in teams of four students to a team.

1. One pair of students for each team selects profile A of Figure O-1, the other pair selects profile B.

2. Each student pair constructs one half of the complete ocean-floor profile from the 40 data points provided for the profile chosen. The data points in Table O-2 (page 87) represent soundings from a ship traverse across the ocean (by the technique described in the preceding experiment).

3. Profiles constructed by student pairs of each team should have the same vertical and horizontal scales in order to match the halves. Student teams can determine the scale they will use after analysis of the data, or the teacher can suggest a scale so that each half of the profile is about 5 cm wide and 30 to 40 cm long.

4. After completion of each profile, the two halves should be joined at the section where overlap occurs. (See Figure O-1.)

ANALYZING RESULTS

The profile constructed is representative of the Trans-Atlantic profiles reproduced in Figure O-1. The student-made profile is constructed from 80 data points or about 1 sounding per 75 km, whereas the Trans-Atlantic profiles in Figure O-1 are constructed using 1 sounding per 1.5 km. Thus, the student-made profiles illustrate topographic features of considerable width and height.

Emphasis should be placed on student recognition of the major topographic divisions, symmetry of divisions around the Mid-Atlantic Rift, and the realization of considerable relief—peaks, ridges, and cliffs preserved unchanged rather than being modified by agents of erosion—on the ocean floor.

Attention can be focused on these principal ideas by questions such as the following:

(1) What happens to the continents at the margins of the ocean?

(2) What changes in slope take place across the profile?

(3) Where do the greatest changes in slope take place?

(4) Are any features arranged symmetrically around a central axis? Which ones?

(5) If more data points were available and plotted, how might the profile change?

(6) If a series of profiles were compared, what similarities and differences would exist?

(7) Compare your profile with other student profiles. If the features change, in what way?

During discussion of these questions it would be advisable to have an overhead transparency of the Trans-Atlantic profiles made from Figure O-1 so that different profiles can be viewed and the effects of vertical exaggeration can be explained.

INVESTIGATING FURTHER

Encourage students to find out how oceanographers, using profiles made from many soundings, perceive the complete ocean-floor topography.

Another problem facing oceanographers concerns the Mid-Atlantic Ridge and its rift valley; what is the significance of the symmetry of major oceanic features around this central rift and what is the origin of this central rift? Could the ocean floor be spreading?

Other avenues of investigation that could be pursued are: how knowledge is gained about the types of sediments and rocks which are found on the ocean floor, the structure beneath the ocean floor, and the possible origin of the topographic features on the ocean floor.

3

Matching Continents
Across the Atlantic
Ocean

MOTIVATORS

Look at a map or globe of the world and note the general shapes of the five continents bordering the Atlantic Ocean—Africa, Europe, Greenland, North America, and South America. Do the shapes of the edges of Africa and South America facing each other suggest that they were once one large continent? Compare also the facing edges of Greenland and North America, and of North America and Europe.

If we assume that these five continents were once combined into one large continent, what might this megacontinent look like? How could these five continents be arranged to form a megacontinent?

BACKGROUND INFORMATION

For about a century, scientists have debated the origin of the continents and ocean basins. Some scientists have supported the theory that the earth has been rigid throughout its history, with fixed ocean basins and continents. Others, bolstered by observing the close "fit" of the facing edges of continents such as Africa and South America, set out to prove the idea that the earth is slightly plastic and that the continents have slowly drifted apart.

For most of this century, the continental drift hypothesis had been in general disfavor among scientists in the United States. However new evidence, provided by the study of terrestrial magnetism, has popularized the continental drifters' position within the past decade. These investigations are based on the fact that many rocks are weakly magnetized at the time of their formation—during cooling of lava or magma (igneous rocks) and during deposition of sediments (sedimentary rocks). The magnetic polarity of the tiny magnetite crystals in these rocks is aligned with the direction of the earth's magnetic field at the place where the rocks formed. The measurements of these polarity directions, called *paleomagnetic readings,* show that these orientations do not fit the present magnetic field at the places where the rocks are now found. This, of course, suggests that the rocks have been moved to their present positions. Furthermore, the magnetic orientation of the rocks on any one continent appears to have changed with the age of the rocks–and in a consistent trend. Those from other continents show different shifts. Continental drift offers an explanation of these findings.

Rocks of oceanic islands provide another bit of new evidence which may support the continental drift hypothesis. If the relative positions of the continents and ocean basins have been fixed, the ocean basins should be generally as old as the continents. However if drift occurred, some regions of the ocean floor could be distinctly younger. Interestingly, age dating (by radioactive isotope decay methods) has disclosed numerous oceanic islands with very young rocks—less than 150 million years old— in comparison with much greater ages for crustal rocks of the continents.

Another line of study is deep ocean-floor sediments. Oceanographers are concerned as to why, in the total span of the earth's history, only a thin layer of these sediments has accumulated. The present rate of oceanic deposition, if extrapolated into geologic time, would extend the process of sedimentation back only 100 to 200 million years. In contrast, the continental history goes back at least 3 billion years.

The foregoing evidences illustrate the type of information that has caused a renewed interest in continental drift. Additional discussion of this hypothesis is included in Oceanography Experiment 4.

MATERIAL

- Map of world (Mercator projection) or world globe

- Partial world map (Figure O-2)
- Scissors

STUDENT PROCEDURE

1. Observe a world map or globe and discuss the relative shapes of the continents across the Atlantic Ocean.
2. Take the copy of the partial world map (reproduced from Figure O-2) and compare it to a classroom map or globe. What similarities and differences can you recognize? (Note that Figure O-2 includes the outlines of the edges of the continental shelves. Decide whether, in attempting to arrange the five continents into one megacontinent, it would be best to include the continental shelves or not.)
3. Cut out the five continents and arrange them in various ways to form a megacontinent. Choose either sea level or the continental shelf edge to cut along.

ANALYZING RESULTS

Compare the various patterns of continent matching which have been devised. Question the students as to which patterns are most logical. (For example it would be more logical simply to slide Africa and South America against each other than to turn either continent around 180°.) The following questions might be raised:

(1) The true border of the continent is better represented by the edge of the continental shelf than by sea level on the continents. Hence, would not the edge of the continental shelf be most representative of the shape of the continental border?

(2) How can areas of overlap of continents when they are joined together be explained? The biggest overlap is Central America which was removed before the match shown in Figure O-3 (page 94) was made; apparently it did not exist there before drifting.

(3) How can areas of gapping (space) between continents when they are joined together be explained? Figure O-3 has a gap in the Gulf of Mexico. Could Central America have come from there?

(4) If the matches of continental borders are reasonably good, is it scientifically fair to conclude that this is simply a coincidence and that the continents were not once joined?

Then compare the student patterns to the fit proposed by some scientists as shown in Figure O-3.

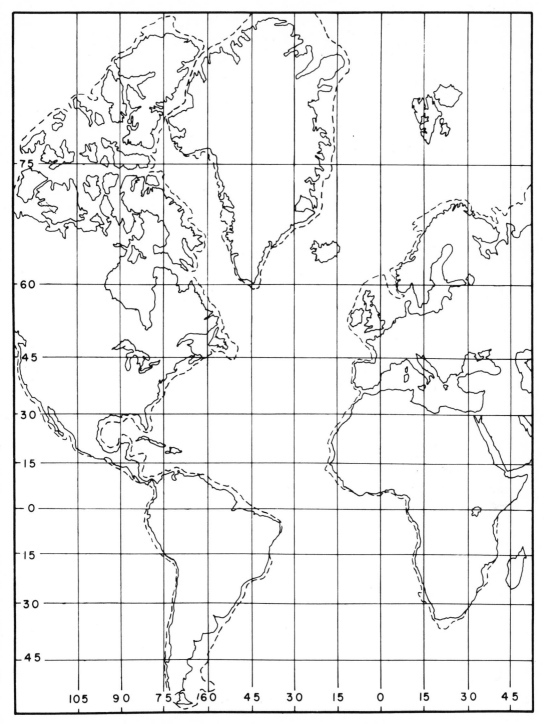

Figure O-2. *Partial world map (mercator projection) showing present distribution of continents adjacent to the Atlantic Ocean. Dashed lines mark approximate outer edge of continental shelf.*

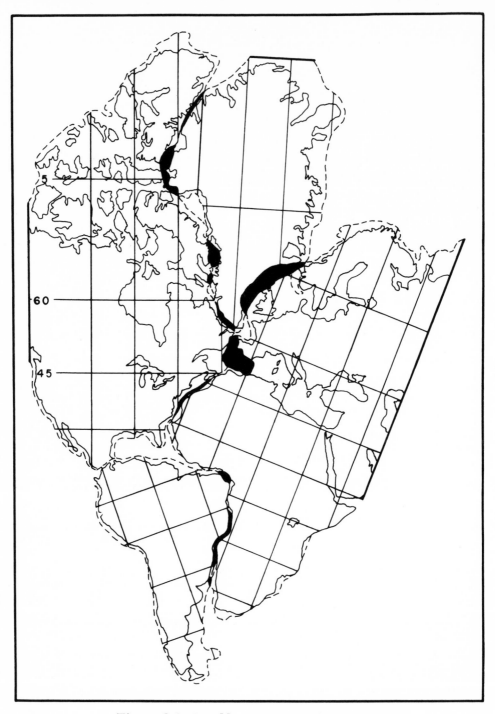

Figure O-3. *Possible orientation of continents as one large continent (perhaps 300 million years ago) prior to continental drift and spreading apart of the Atlantic Ocean sea floor. Fit along outer edge of continental shelf (dashed lines). Areas of overlap shown in black; gaps indicated as blank areas between dashed lines.*

INVESTIGATING FURTHER

Looking at the world map or globe, what shifting of the other continents (Antarctica, Australia, Europe-Asia) might have taken place? Are there any areas that appear as "openings" that could be readily closed by sliding continents together? Be sure to look carefully at the Mediterranean Sea-Black Sea area as one example. By shifting Saudi Arabia south, could the Red Sea and Gulf of Aden be readily closed?

Some oceanographers and geologists support the idea that India has moved northward under the Tibetan Plateau (and caused the folding and uplift of the Himalayas). By moving India south, the Himalayas would be "smoothed out" and the Arabian Sea closed.

Suppose the continents have shifted from former positions, forming the North and South Atlantic and Indian oceans as they spread. Could we expect to see any evidence of this spreading in these oceans? Where might we look? How about near the *centers* of the oceans. Interestingly, we find ridges in the centers of these oceans, so let's examine them next.

4

Ocean Ridges–
An Expanding Earth?

If we accept the hypothesis of continental drift—the splitting up of a megacontinent and the fairly constant drifting apart of the pieces—where would a likely place be to look for evidence of this movement? What kinds of evidence would you look for? What might happen to the earth's crust as the megacontinent split up? And, what would cause the continents to move apart?

BACKGROUND INFORMATION

Theories in science live or die by the results of being tested. No matter how attractive an hypothesis is, unless it agrees with observation and examination, it cannot be taken seriously.

A brief look at the history of the continental-drift hypothesis demonstrates the extraordinary complexity of this subject. A German meteorologist, Alfred Wegener, felt the need to explain the parallel development of living things throughout the world. Similar fauna and flora have existed in widely separated regions throughout geologic history. A common explanation was the supposed existence of land bridges linking the various

96

rigidly fixed continents, however no evidence exists for most of the supposed bridges.

In 1910, Wegener suggested that once a single, giant continental mass called *Pangaea* existed, and that the rest of the earth was covered by an ocean called *Panthalassa*. In time Pangaea split up and the pieces drifted apart as the continents of today.

Unfortunately for Wegener's hypothesis, at that time no forces strong enough to shove the continents around, let alone split one into fragments, were known to exist. A first lead to understanding such a mechanism came from measurements of variations in magnitude of the earth's gravitational field. These measurements revealed the presence of major trenches, best explained by subcrustal movement of material within the earth's mantle, pulling down the ocean floor. Earth scientists have proposed the existence of "convection cells" moving by heat transfer as the cause of these subcrustal movements.

Discovery of regions where these currents appear to ascend toward the earth's surface is perhaps the strongest confirmation of this hypothesis. The discovery of such regions represents a major accomplishment in recent ocean-floor explorations. These topographic features, as grand in scale as the continents themselves, are known as *mid-ocean ridges*. Measurements of the flow of heat outward from the earth's interior have revealed high values along the ridges. In fact, the values are unusually great, exceeding by two to eight times the average flow observed on the continents and elsewhere on the ocean floors. In contrast, measurements show that the flow of heat in the oceanic trenches falls to as little as $\frac{1}{10}$ the average value.

Through the study of terrestrial magnetism, oceanographers and geophysicists have observed that the earth's magnetic field has not only changed direction slightly, but also has universally reversed polarity numerous times in the past. By first dating samples of these rocks (using radioactive decay methods to establish their radiometric ages), and then carefully determining the magnetic polarity of these same samples, scientists have established that polarity of the magnetic field was reversed at certain fixed times in the past. That is, the north-south magnetic orientation was reversed from the present position. Such reversals could be explained by a *change* in the differential motion between the earth's core and mantle from the present motion. (The differential core-mantle motion is likened to a dynamo, and believed by many physicists to cause the earth's magnetic field.)

By exploring the ocean floor using a magnetometer (an instrument

97

for measuring magnetic intensity), oceanographers have found the orientation of magnetic polarity to occur in bands over large areas forming a stripelike pattern. Interestingly, this pattern of stripes is symmetrical to the mid-oceanic ridges about centers which are the median rift zones of the ridges.

Incorporating the convection cells, magnetic reversals, ages of reversals, and symmetrical stripelike patterns observed, earth scientists have proposed the hypothesis of ocean-floor spreading (*see* Figure O-4). According to this hypothesis, new ocean floor is continuously created by hot magma rising toward the rift in the center of the ridge. It becomes magnetized in the direction of the earth's magnetic field as it cools, and then moves outward laterally from the ridge center as more magma intrudes into the rift from below.

The importance of the relation of ocean-floor spreading to continental drift is apparent. Ocean-floor spreading provides a mechanism, in harmony with physical theory and substantiated by geological and geophysical observations, for disrupting and moving continents by lateral movement of the crust.

However, ocean-floor spreading does not unequivocally call for continental drift. Ocean-floor spreading could occur without drifting continents. Nonetheless, the directions and rates of motion for both are entirely compatible. Above all, a major objection to the hypothesis of continental drift has been removed.

MATERIAL

- Clay (4 colors)
- Plastic knife
- Colored pencils

STUDENT PROCEDURE

1. Shape pieces of different-colored clay into five rectangular sections $\frac{1}{2}$ to 1 cm thick, 5 to 8 cm long, and widths of approximately $\frac{1}{2}$, 1, 2, 3, and 4 cm. (A class decision should be made so that similar-sized sections have the same color.) The clay sections represent magmatic material intruded along the central fracture of the mid-oceanic ridge system.

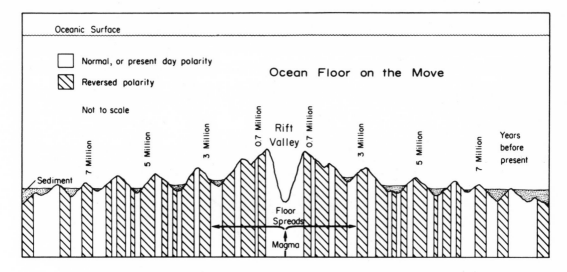

Figure O-4. *Diagrammatic view of a section of the crust across the Mid-Atlantic Ridge showing symmetry of magnetic polarity and ages of the rocks about the central rift valley of the Ridge.*

2. To assemble, choose one clay section and slice it lengthwise representing the central rift of the ridge.
3. Place arrows facing upward on one end of each of the slices.
4. Then part the slices and insert a different-colored section between them. The arrows represent the direction of magnetic polarity of the magma at the time of cooling. The inserted section of different-colored clay represents a subsequent magma intrusion following splitting apart of the ridge (*see* Figure O-5, next page).
5. Mark the new clay section with arrows pointing down (to represent reverse orientation of the magnetic polarity).
6. Slice and part it and insert another clay section.
7. Continue the procedure with each clay section, each time reversing the direction of the arrows from the previous orientation.
8. After completing this assembly, draw a cross-sectional (end) view of the pattern obtained. Use the profile obtained across the Mid-Atlantic Ridge (Oceanography Experiment 2) to guide your drawing. The profile should have a central valley (over the last inserted clay section) and then be ridge shaped in a symmetrical pattern as noted by the profile of Experiment 2. Be sure to place the arrows on the view as marked on the clay model.

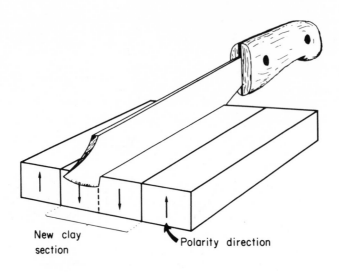

New clay
section

Polarity direction

Figure O-5. *Procedure for slicing clay sections
when making a model of the Mid-Atlantic Ridge.*

ANALYZING RESULTS

Begin by comparing the cross sections drawn by the students with Figure O-4. The analogies of the model to the hypothesis of mid-ocean ridge development should be carefully discussed.

The stripelike pattern of colored clay resulting from insertion of each section (intrusion and cooling of magma) along the central slice (rift valley) should be emphasized. The colors do not represent different rock types, but rather different ages of rock (youngest in center and progressively older going out in each direction). The arrows signify polarity directions. Note the symmetry of both colors (ages) and arrows (polarity directions) around the mid-oceanic ridges.

Further thought and clarification of ideas can be obtained by discussion of questions such as the following:

(1) How do oceanographers determine the ages of the zones of different-aged rocks with different magnetic polarities?

(2) Oceanographers have established that the age of the stripelike magnetic polarity increases symmetrically away from the center of the ridge (Figure O-4). What might this suggest about possible movement of the ocean floor?

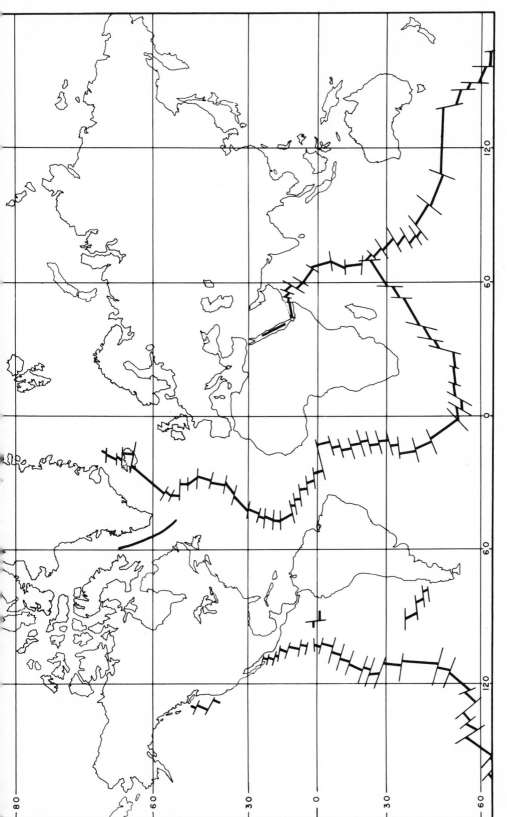

Figure O-6. *World map (mercator projection) showing generalized pattern of the crests of oceanic ridges (heavy lines) and transverse fracture zones (light lines) cutting across the ridges.*

(3) Oceanographers have recorded high heat-flow values over the centers of mid-oceanic ridges (the rift-valley areas) compared to lower values off the ridge flanks and over other parts of the oceans and the continents. What might this suggest about the circulation pattern of convection cells within the mantle?

(4) Might the width of the mid-oceanic ridges be an indication of the amount of spreading of the ocean floors?

INVESTIGATING FURTHER

On a map of the world (mercator projection) place the mid-oceanic ridges where you think they would occur. Remember: they are generally *mid*-oceanic! Now compare your map with Figure O-6 which has these ridges located on it.

Note the offsets of the mid-oceanic ridges, especially on the Mid-Atlantic Ridge, caused by nearly east-west trending faults. (These faults commonly follow latitude directions.) How might these faults have been caused? (They may be related to local movements of the ridges due to magma intrusion concentrated in local areas.)

By estimating the total area of the mid-ocean ridges around the world, would it be feasible to estimate a possible total expansion of the earth—if, indeed, the earth has expanded? (This approach would probably be fruitless because the ridges could be formed without total global expansion.)

If the ocean floor is spreading, we would expect it to either shove against the edges of the continents or slide down under the continental edges. Looking at a relief map or relief globe of the world, do you see any evidence for this shoving or sliding of ocean crust under the edges? (Attention should be given to the belts of folded mountains, such as the Andes, Appalachians, and Rockies, which commonly lie along the margins of the continents. These mountains may reflect compressional folding in response to lateral shoving of sea-floor material against the continents.)

5

Shoreline Studies
from Highway and
Topographic Maps

MOTIVATORS

The Pacific Coast of the United States is characterized by a relatively straight coastline, rugged topography, and high relief. In contrast, the shoreline along the eastern U. S. is highly irregular with many embayments, and coastal relief is low. How can the differences between these two ocean-land *interfaces* be explained? Are not the same erosional and depositional processes at work along both coastlines? And how does this ocean-land interface relate to man's livelihood and his population concentrations?

BACKGROUND INFORMATION

Coastal features result from many different agents and processes. Changes of sea level associated with Pleistocene glaciation, erosion and deposition by waves and currents, crustal movements, and organic buildups have all left their mark upon the shoreline.

Major coastal features and their modes of origin are:

Bay and estuary. Most indentions along coasts owe their existence to submergence that results from rising of the seas or sinking of the land.

103

If the previous land surface was erosional, as is generally true, the shore-line resulting from drowning will be irregular because it reflects the previously eroded surface. Embayments resulting from submergence are called "drowned valleys," bays, or estuaries depending on size and shape.

Bar or spit. Sand and gravel move parallel to the shore by obliquely striking waves and longshore currents until a protected position is reached where they are deposited. In time these deposits may build up to, or above, the water surface, forming a point of sand or gravel that extends outward from the shore. Such features are called bars or spits. In a narrow or shallow bay, a spit may grow across the entrance forming a *bay-mouth bar*. In broad bays, the shoreward movement of waves and currents may cause the spit to have a shoreward curve, forming a *hook*.

Barrier beach and barrier bar. Along coasts where shallow water causes waves to break far from shore, a barrier bar develops just inland from the line of breakers. The bar first forms above sea level as a long, narrow ridge. Further deposition by waves and longshore currents develops a series of ridges and fills the gaps between them to form a *barrier island*.

Between the barrier bar and the mainland is a long, narrow, shallow *lagoon*. Drainage from the land discharges into the lagoon and seeks an outlet through the barrier island, and in the lagoon the water level rises and falls periodically with the tide. Thus gaps or *tidal inlets* between barrier bars are kept scoured open by water flowing in and out through them.

Wave-cut cliffs and wave-cut terraces. On coasts where the water deepens rapidly, waves break directly against the shoreline, and thus expend the greatest part of their energy by eroding the land. In this manner they form features called wave-cut cliffs. Wave erosion pushes the wave-cut cliff steadily landward, producing a shallow submerged plat-form called a wave-cut terrace, at its foot. After formation, wave-cut terraces are commonly left high and dry by rising of the land associated with diastrophic activity.

These cliffs and terraces are commonly found many hundreds of feet above present sea level. Some coasts display a series of such terraces, forming giant steps in the profile of the land as it rises from the present shore.

Delta. Where a river enters a body of relatively calm water, a lake, or the ocean, its velocity and transporting powers are quickly reduced. Deposition of the material it carries will build a delta. Not every river has a delta. Some transport little sediment; others deposit their load on coasts with deep water or expose it to violent wave action so that the sediment is spread over the adjacent sea floor, or longshore currents transport the sediment laterally.

MATERIAL

- Highway map of coastline near your locality [1]
- Topographic maps of coastline near your locality, if available [2]
- Aerial photograph pairs [3]
- Stereoscope [3]

STUDENT PROCEDURE

Regional highway maps and, if available, topographic maps of shoreline areas near your locality should be provided for the class. Divide the class into teams of two members each.

1. Locate and identify shoreline features and determine what erosional and depositional processes influenced their development.
2. After briefly studying the maps in this manner, select one of the following sections of coastline:
   A. EAST COAST
      1. Charleston, South Carolina to Jacksonville, Florida.
      2. Atlantic City, New Jersey to North Carolina-Virginia state line.
   B. GULF COAST
      1. Gulfport, Mississippi to Port Arthur, Texas.
      2. High Island, Texas to Corpus Christi, Texas.
   C. WEST COAST
      1. Columbia River mouth, Oregon-Washington state line to Eureka, California.
      2. San Francisco Bay entrance, California to San Diego, California.
3. Using highways maps, trace these coastlines on a sheet of paper. (Be sure that each team has one West-coast section and one East- or Gulf-coast section. (See Figure O-7, next page.)
4. Upon completion of outlines, compare the similarities and differences of coastal features of the West-coast sections with the East- and Gulf-coast sections.

---

[1] The use of a regional highway map is recommended. These maps are readily attained at most gas stations.
[2] Refer to U.S. Geological Survey series of topographic maps. See Appendix V.
[3] See Aerial Stereo-Photographs by Hubbard Scientific Company (Appendix V).

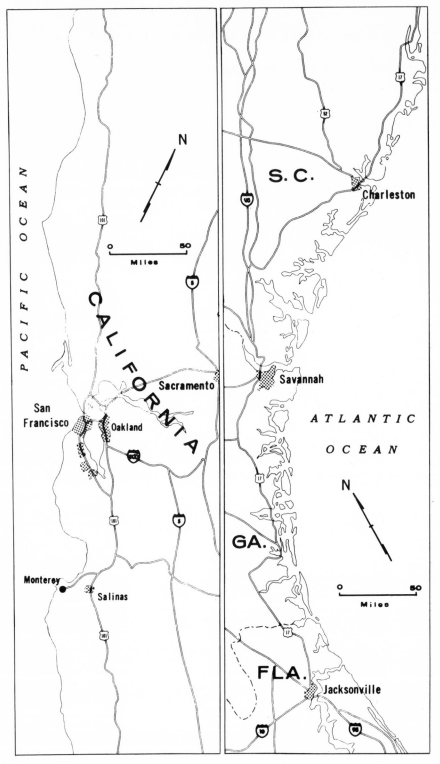

Figure O-7. *Segments of highway road maps showing relatively straight coastline of California compared to highly irregular coastline from South Carolina to Florida.*

TEACHER NOTE: If available, provide the student teams with aerial photograph pairs and topographic maps representative of shoreline features located along the East, Gulf, and West coasts of the U. S. These materials will enable the students to visualize coastal features in three dimensions, determine relationships of each shoreline feature to surroundings, including: relief, degree of ruggedness, and overall shape—all of which are necessary in making meaningful comparisons.

Questions, such as the following, may be used to direct student observations of similarities and differences that exist along the coastlines.

(1) How do the different sections of coastline drawn compare, especially in regard to a contrast of shoreline regularity?

(2) What does the number, size, and drainage pattern of major rivers indicate about the topography landward of the coastline?

(3) How does the topography seen affect the erosional and depositional processes active along the shoreline interface?

(4) What agents of erosion and deposition are active in shaping and changing shorelines? Which do you believe are most effective?

(5) How would the degree of ocean-floor slope affect the erosional and depositional agents active along the shoreline interface?

(6) How and why does the shape and arrangement of shoreline-depositional features indicate the direction of current movement and wave action?

(7) Why are large depositional features, such as deltas, not located along the West coast?

(8) What effect has shoreline topography had on the distribution of population centers and the development of major highway systems?

ANALYZING RESULTS

Based on the above questions and their discussion, have students list the reasons that could account for the major differences—degree of regularity, ruggedness, and degree of relief—in coastline topography between the Pacific coast and the Atlantic or Gulf coasts.

Then discuss these reasons, emphasizing that most shoreline features result from present-day erosional and depositional processes operating on coastlines which were formed either by crustal uplift or crustal sinking.

Certain features, such as estuaries, reflect the drowning of the previously eroded surface producing a highly irregular, low-relief shoreline.

In contrast, elevated wave-cut cliffs and wave-cut terraces reflect the diastrophic uplift that is occurring and results in a relatively straight, rugged shoreline.

Other shoreline features result from longshore current distribution and deposition of sediment being transported by rivers to the shore. Thus features such as bars, spits, hooks, and bay-mouth bars are related to present-day erosional and depositional processes and may be found along any section of the U. S. coastline.

Delta development requires relatively calm water with little wave and current action over an extended period of time. Thus, large deltas— like the Mississippi delta—are not developed along most of the U. S. coastline because of the constant interplay of oceanic waves and currents. However, numerous small deltas develop where sediment-laden rivers enter bays, estuaries and lagoons, and often landward of a tidal inlet.

Have the class consider the following question: Would the colonization and subsequent industrialization of the U. S. have been affected if the first explorers had come upon a coastal environment similar to West-coast topography rather than the deeply embayed Eastern coastal plain topography? Embayed shorelines provide the natural harbors and the calm waters needed for safe anchorage of ships and subsequent loading and unloading of cargo. Deep landward penetration by estuaries offers the advantage of allowing harbor facility development near the areas where food and goods are produced or raised. Rivers flowing on coastal plain topography are usually navigable a considerable distance upstream, allowing an easier exchange of cargo and movement of people between coastal and inland regions.

INVESTIGATING FURTHER

Using regional highway maps, have students observe the shoreline around the Great Lakes. Have them locate and identify as many "ocean shoreline" features as possible. Then ask them to determine the direction waves were moving when the features formed. They should also be able to infer the direction of wind which produced the waves because of the relatively small size of these lakes when compared to the oceans.

If time and resources are available, a field trip to an ocean shoreline would provide the students with a firsthand experience of the processes that interact to form shoreline features. A lake shoreline will also have many features common to oceanic shorelines, although on a somewhat smaller scale.

Are all coastline features the result of wave and current interplay with mountainous and coastal-plain topography? Have students observe coastal shorelines along the major continents of the world. Is their any evidence to suggest that other agents may influence shoreline development? How about wind? Many shorelines have dunes formed by wind transport and deposition of sand-sized material. What about glacial ice? In many localities, glacial ice is actively shaping the shoreline topography.

What effect did continental ice sheets have on shorelines during the "ice ages"? Numerous ice-scoured embayments called *fiords* are found along the coastline of northeastern U. S. and on other shorelines.

6

Shoreline Studies from Aerial Photographs

Imagine that you are in a spacecraft and through its window you see the view illustrated by Figure O-8 (page 113). One of the first features you would notice is the meeting place of ocean and land—the shoreline. Have you ever wondered why some shorelines are nearly straight while others are highly irregular and deeply embayed? Would waves that continuously break onto the beach, or sediment that is carried to the oceans by rivers, have anything to do with the appearance of a shoreline? Aerial photographs may provide clues to answer these questions.

BACKGROUND INFORMATION

The energy that works upon and modifies a shoreline comes largely from the movement of water by wind-formed waves, from tides, and, to a lesser extent, from sea waves, called *tsunamis,* generated by earthquakes.

Wind-formed waves and associated coastal currents are the most important processes continually reshaping the shorelines.

The area between the shoreline and the outermost limit of breaking waves is known as the *surf zone.* The fall of water in a breaker produces the wave's greatest erosional effect on the shallow ocean floor. As a wave breaks, not all of its energy is expended on erosion of the shoreline. The angle at which the wave front approaches the shore results in a flow in the surf zone along the shore called a *longshore current*. After flowing parallel to the beach, the water is returned seaward in a narrow flow called a *rip current*.

Most waves advance obliquely toward a shoreline, thus as a wave front nears the shore, the section closest to land feels the effect of the bottom first and is retarded, while the seaward part continues shoreward at its original velocity. This effect, known as *refraction,* causes the wave to change direction and approach the shore nearly head-on. When waves pass over submarine ridges extending out from the headlands and submarine depressions extending out from the bays, refraction causes the wave fronts to converge over the ridges and diverge over the canyons, thus concentrating the energy on the headlands.

In addition, the attractive forces that operate between the sun, the moon, and the earth set the waters of the oceans in horizontal motion to produce *tidal currents*. Currents due to tides and their effect on shoreline features are in general subordinate to those of wind-formed waves. However, if local conditions are favorable, these currents can be important in shaping and maintaining shoreline features. The swiftest currents usually build up where a body of water—lagoon or bay—has its only access to the open ocean through a narrow, restricted passage. These passages, called *tidal inlets,* are probably first opened by violent storms, possibly of hurricane force, and are then maintained by the periodic rise and fall of the tide and its associated currents.

Shoreline features are never stagnant, and are only part of an ever changing, dynamic system dependent upon the available energy from wind-formed waves and tides.

MATERIAL

- Aerial photograph (Figure O-8 or similar photograph)
- Aerial photograph pairs [1]
- Ripple tank or stream table [2] (if available)
- Highway map from Oceanography Experiment 5
- Magnifying glass

---

[1] See Aerial Stereo-Photographs by Hubbard Scientific Company (Appendix V).
[2] See *Geology and Earth Sciences Sourcebook* (Appendix V).

STUDENT PROCEDURE

Provide the students (in small groups) with photographs showing a view along a shoreline. (Figure O-8 is a good example.)

1. Construct a sketch map of the area with the outline of the shoreline as was done in Oceanography Experiment 5. The sketch map should include all patterns shown in the water as well as features on land.

2. Compare this map with the shore outline map constructed in Experiment 5 for similarities and differences.

3. Locate and identify on the aerial photograph the shoreline features discussed in Experiment 5 and determine what additional features can be seen that were not evident on the highway map. (A magnifying glass will be helpful in determining features on the aerial photograph.)

Questions, such as the following, may be used to direct observation of the aerial photograph features:

(1) Why does the water have light- and dark-colored areas rather than being one continuous shade? (As seen in Figure O-9, page 114, the light-colored zones are sediment-laden waters closely associated with discharge from the bays and lagoons.)

(2) Can you recognize the direction of prevailing winds and longshore currents?

(3) What depositional features (bars, spits, etc.) can you recognize?

(4) What characteristics of the land surface can be identified (e.g. drainage pattern, nature and density of vegetation, features of the bedrock, etc.)?

(5) What man-made features can you locate on the aerial photograph?

If available, provide the student teams with aerial photograph pairs showing representative shoreline features. A three-dimensional view of beach ridges, sand bars, tidal deltas, and barrier islands will enable the students to see the relationship between shoreline sediment transport and shoreline depositional features.

TEACHER NOTE: Refraction of waves as they approach the shore and concentration or dissipation of wave energy on the shore can be demonstrated with the use of a ripple tank. Wave and current effects along shorelines, development of coastal terraces by wave action, and the effects of offshore slope can be demonstrated by a stream table. For information on construction and demonstration procedures consult the *Geology and Earth Sciences Sourcebook* (Appendix V).

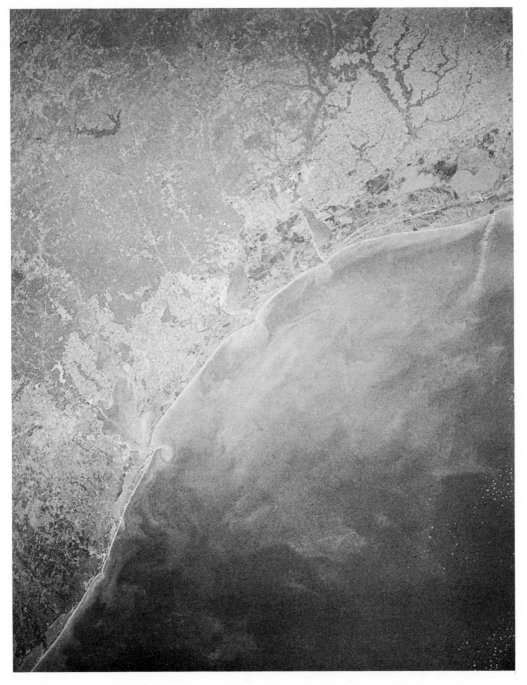

Figure O-8. *Photograph of Texas and Louisiana coast taken on Gemini flight XII, November 14, 1966 by astronauts Capt. James A. Lovell, Jr. and Lt. Col. Edwin E. Aldrin, Jr. (NASA/MSC No. 566-63184; courtesy of National Aeronautics and Space Administration).*

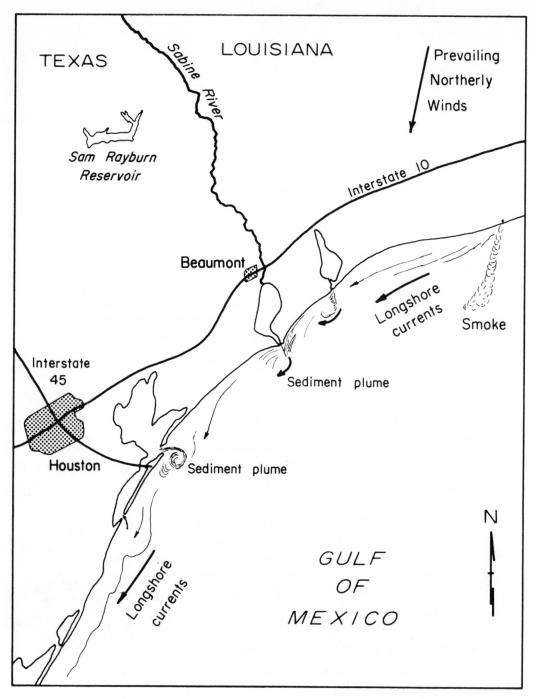

Figure O-9. *Sketch of photograph shown in Figure O-8 identifying geographic and cultural features and showing pattern of sediment discharge and longshore currents.*

ANALYZING RESULTS

After a discussion of the above questions, have students list the conditions necessary for the development of shoreline topography illustrated in the aerial photograph. Discuss these conditions, emphasizing that the shoreline appearance is commonly the result of sediment transported by longshore currents and subsequent deposition along a shallow, gently sloping ocean bottom.

Many features observed in Figure O-8 reflect the present-day sediment-transporting processes operating along other shorelines. Light-colored swirls, located seaward of inlets, indicate a mixing of sediment-laden waters from the land with the longshore movement of ocean waters. Many of these swirls have an elongation parallel to the shoreline, reflecting the general southwesterly movement of the longshore currents of the region. Lighter-colored areas, elongated perpendicular to the shoreline, reflect seaward movement of sediment by rip currents or discharge.

Figure O-8 was taken when the wind was blowing from the north. (Note the direction the smoke is moving from a marsh fire along the coast.) The brisk northerly winds account for the southwestward-flowing longshore currents and the seaward movement of water from the lagoons and estuaries.

Roads, ship channels resulting from deepening inland bays for ocean-going ships, and the Intracoastal Canal located parallel and landward of the barrier islands are some of the man-made features recognizable in Figure O-8.

Clouds are located in the lower right-hand portion of the photograph and their shadows can be observed on the ocean surface. The shadows indicate the sun's position to be slightly south and almost directly above the spacecraft.

INVESTIGATING FURTHER

If available, provide the students with aerial photographs of shoreline areas of various parts of the United States. The contrast (as described in Oceanography Experiment 5) between different shorelines is dramatically shown on photographs. These contrasts could be well demonstrated by a comparison of the East coast, Gulf coast, and West coast as outlined in Experiment 5.

For those fortunate enough to live near the coasts, a field trip to an ocean shoreline would provide the students with firsthand experience about the erosional and depositional processes operating within the surf zone.

Other possible avenues of investigation include man's effect on the natural processes of longshore transport and deposition of sediments, the chemical composition of lagoonal waters and its ecological effects, and the economic and recreational importance of shorelines. Construction of barriers along the shoreline (such as breakwaters, groins, and jetties) and dams on the rivers flowing into the oceans may have a significant effect on nature of sediment transport and deposition along the shore.[3]

---

[3] An excellent film illustrating the effect of man's involvement in the natural pattern of sediment transport is *The Beach—A River of Sand* by the Encylopaedia Britannica Film Corporation. See Appendix V.

116

7

Pressure in the Ocean–
What Affects It?

MOTIVATORS

Can you imagine the conditions that exist on the ocean floor? Because light does not penetrate very deeply into water, it is very dark. Thus most life cannot exist there. And because the ocean floor is "buried" by many hundreds of meters of water, the pressures are very high. Even submarines operate at relatively shallow depths because at great depths the high pressures would cave in their walls.

BACKGROUND INFORMATION

Pressure is the amount of force produced on a unit area. In any fluid, such as air or water, this force is caused by the pull of gravity on the fluid above the point at which the pressure is measured.

The pressure exerted on any fluid is thus affected by two variables— the *nature* of the fluid and the *depth* of fluid. Therefore, the pressure a few inches below the surface in a lake (of fresh water) differs from that at the same depth in an ocean (sea water). And likewise, the pressure a few inches below the water surface is less than that a few feet below the surface.

To test independently each of these variables, experiments must be performed changing only one of these conditions at a time. Using only tap water, the changes in pressure with changes in depth can first be evaluated. Then, at constant depth, the significance of the nature of the fluid can be determined by using different liquids.

MATERIAL

- Glass tubing bent into a U-shape, and mounted on a board (pegboard is convenient)
- Rubber tubing (approximately ½ to 1 m length)
- Beaker (large)
- Thistle tube
- Balance
- Liquids: Alcohol, cooking oil, mineral oil, liquid soap, and sea water.

STUDENT PROCEDURE

Part A.

1. Begin by adding colored water to the U-tube so that it is about half full. Note that the water level is equal on both sides of the U-tube.
2. Attach the rubber tubing to one side of the U-tube and have a student blow into the tube with his mouth near (but *not* on) the tube. What happens to the level of the water in the U-tube?
3. Place the end of the rubber tube just below the surface of the beaker full of water and observe what happens to the water level in the U-tube.
4. Move the end of the rubber tube up and down in the water and notice how it causes the water level in the U-tube to fluctuate (see Figure O-10). To accentuate this effect, attach the thistle tube to the end of the rubber tube and then move it up and down in the water.
5. Place a scale along one side of the U-tube as shown in Figure O-10. Record the level of water in the U-tube as the depth of the thistle tube is increased by units of 1 cm (See Table O-3, page 120).

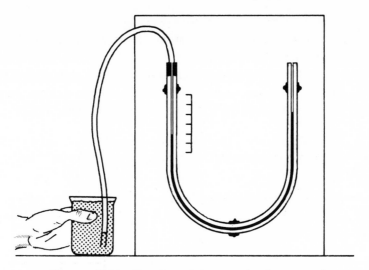

Figure O-10. *Sketch showing technique for measuring change in pressure with depth in container of tap water. Note scale on pegboard to measure fluctuations of liquid in the U-tube.*

Part B.

You have now seen the effect of depth changes of the thistle tube in one liquid—tap water.

6. Compare the results with those using other liquids, for example alcohol, cooking oil, mineral oil, liquid soap, and sea water. For each, place the thistle tube the exact same depth below the surface of the liquid (for example 3 cm) and record the water level in the U-tube on a data sheet.

7. To understand better why the pressures differ for each liquid, carefully weigh 100 ml of each liquid and record these weights on a data sheet.

ANALYZING RESULTS

Part A.

The fluctuation in water level in the U-tube is a response to the difference in pressure between the open end of the U-tube (being pressed by the air) and the end of the thistle tube (being pressed by the depth of water). Thus the change in height of the water level in the U-tube is a measure of pressure changes (in centimeters of water) as the thistle tube is moved down in the water.

Determine the relationship between pressure and water depth by graphing the results (see Table O-3). Note that the change in pressure increases in constant proportion to the increase in water depth.

Table O-3. *Data sheet and graph for recording changes in pressure with changes in depth of tap water.*

Depth of thistle tube below water surface (in cm)	0	1	2	3	4	5
Pressure: change in level of water in U-tube (in cm)						

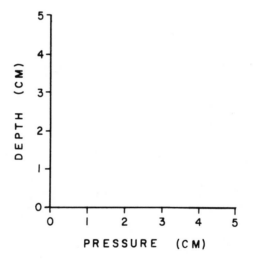

Part B.

You have demonstrated that at the same depth different liquids exert different pressures. If this is a response to the "pull of gravity," the weight of the liquids is significant. Heavier liquids will exert more force per unit area because the pull of gravity is directly affected by the mass of liquid being pulled. To establish further the relationship between weight and pressure, prepare a graph of these values, which are shown on the data sheet accompanying Table O-4.

Table O-4. *Data sheet and graph for recording weight and changes in pressure of different liquids measured at constant depth of thistle tube below water surface.*

Liquid	tap water	alcohol	cooking oil	mineral oil	liquid soap	salt water
Weight of 100 ml (in gm)						
Pressure: change in level of water in U-tube (in cm)						

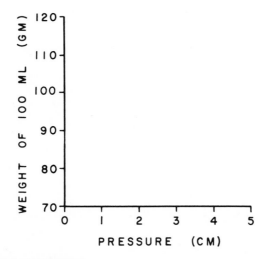

INVESTIGATING FURTHER

Sea water has a density of 1.03 g/cm³ as compared to 1.00 g/cm³ for fresh water. The pressure 1 cm below the water surface is thus 1.03 g/cm² in the ocean; 1 m below the water surface it is 103 g/cm². What would be the pressure at 1-km depth in the ocean? How does this compare with a 1-km depth in a fresh-water lake?

Because rock is considerably heavier than sea water, pressures developed by burial under rock are proportionately greater than under equal depths of sea water.

The average density of crustal rock is 2.7 g/cm^3. Thus a 1-cm cube of rock pushes downward with a pressure of 2.7 g/cm^2. Therefore, rock buried 1 m below the surface is at a pressure of 270 g/cm^2. What would be the pressure on rock at a depth of 1 km?

It is, therefore, not difficult to estimate the great pressures that must exist in the earth's lower crust and mantle. At these depths, it is likely that the rock (because it is at high temperatures as well as high pressures) behaves more like a fluid than a solid. This might help to explain how continental drift and ocean-floor spreading could happen.

8

What Makes Sea Water Heavy?

MOTIVATORS

Many earth scientists support a "degassing process" during some stage of the earth's formation as a means of deriving many of the elements common to the atmosphere, oceans, and to life itself. Whether this release of matter—especially water, carbon dioxide, and chlorine—took place rather abruptly at an early stage in the earth's development or has been a gradual process through geologic time is not known. However, the chemical composition of the oceans has not remained constant. Material derived from weathering of rocks on the continents is continually being added by streams. In this manner, appreciable amounts of dissolved materials including calcium, magnesium, potassium, silica, and sodium have accumulated in the oceans.

BACKGROUND INFORMATION

Sea water is heavier than tap water; this was observed in the preceding experiment. But what other distinguishing features characterize sea water? One answer that immediately comes to mind is that sea water is *salty*. The distinctive taste is, of course, caused by the presence of salts in

123

solution. These materials add to the overall weight of sea water and account for the difference in weight of equal volumes of sea water and fresh water. The amount of dissolved salt in sea water can be readily measured by the method described in the Student Procedure.

MATERIAL

- Balance
- Bunsen burner
- Evaporating dishes (2)
- Ring stand and wire gauze
- Sea water

STUDENT PROCEDURE

1. Carefully weigh the two evaporating dishes.
2. Add tap water to one dish and sea water to the other. Be sure that equal amounts of each (by volume) are used.
3. *Slowly* heat each dish until all the water is evaporated.
4. Reweigh the dishes and determine the increase in weight of each. The difference between this weight and the first weight of each dish represents the amount of mineral matter that was dissolved in the water prior to evaporation. Note that the amount is greater in the sea water.
5. Next determine the per cent of material dissolved in sea water. To do this begin with a known amount (by weight) of sea water. For example, begin with 15 g of sea water.
6. Place it in one of the evaporating dishes of known weight.
7. Evaporate the water as before.
8. Then reweigh the dish and determine its change in weight—the amount of mineral matter that had been in solution.
9. Figure the per cent of mineral matter contained in the sample by dividing the weight of mineral matter by the original weight of sea water (15 g).

ANALYZING RESULTS

The mineral matter remaining in the dishes after the water was evaporated represents material normally in solution. Note that some mineral matter occurs in the dish which contained tap water. However, this amount is significantly less than that contained in sea water. These dissolved materials largely accumulate by chemical weathering of rocks on the continents. Since water generally moves over the continent in a relatively short period of time, the salts dissolved by weathering are not concentrated in the streams. In contrast, water flowing over the continents is continually carrying dissolved salts into the oceans. Water is removed from the oceans by evaporation, leaving the salts in ever increasing concentrations there.

The concentrated salts dissolved in sea water give it the salty taste and the added weight when contrasted with fresh water.

INVESTIGATING FURTHER

Have the class discuss the problem of what limitations are presented in calculating the total age of the oceans, based on the annual *rate* of salinity increase compared to the present total salinity of the oceans. Assuming no salt content in the original oceans, it is a simple problem of comparing annual rate to total percentage—or is it?

Perhaps the greatest difficulty involved is that dissolved mineral matter is continually being *extracted* from sea water. The abundant deposits of the sedimentary rock, limestone, attest to the amount of calcium carbonate which is precipitated out of the oceans. Salt too is frequently removed from sea water, as evidenced by extensive deposits of salt now found buried with marine sedimentary rocks. It would be necessary to calculate the total volume of precipitated salt now present in sedimentary rocks of the crust and add this to the amount of salt presently in solution in the oceans before a meaningful value of total salts could be obtained. What other considerations would have to be made before a meaningful figure was assured? (One would have to be able to estimate with a fair degree of accuracy the amount of salts recirculated—that is, once deposited in the oceans then precipitated out to form sedimentary rock, and later weathered and eroded back to the oceans.)

9

Measuring Density
Differences of Water

MOTIVATORS

W hat effects do the additional materials contained in solution have on sea water? At similar depths, the pressure in sea water is greater than that of fresh tap water as shown in Oceanography Experiment 7. Is this why it is easier to swim and float in the oceans than in a lake? When an object floats in water, part is submerged *in* the water, part is *above* the water. If the object is very light, only a small part is in the water and the object "rides high." In contrast, only a small part of a heavy floating object lies above the water—most lies in the water. But what if the object remains the same and the weight of the water changes? Will this cause a change in the floating height of the object?

BACKGROUND INFORMATION

The *density* of a material is the *mass of one unit volume* of the material. Mass is commonly measured by weighing the material; however mass doesn't change, but weight might. For example, if you had accompanied the astronauts who landed on the moon, your weight on the moon

126

would have been much less than it is on earth—about ⅙ your normal weight. But your mass wouldn't have changed, just the gravitational pull on your mass would have decreased, so you would weigh less.

Using a homemade hydrometer like the one illustrated in Figure O-11, the densities of tap water and sea water can be compared. The denser the liquid used, the higher the dropper will float. This is because an upward pressure is exerted on the dropper by the liquid. To float, the dropper must displace enough liquid to equal its own weight. Not as much of the denser (heavier) sea water is needed to support the dropper—so it floats higher! Hence, we can conclude that *the denser the liquid, the higher the dropper will float in it.*

MATERIAL

- Graduated cylinder
- Eye dropper
- Liquids: alcohol, cooking oil, mineral oil, liquid soap, and sea water

STUDENT PROCEDURE

1. Pour tap water into the graduated cylinder so that it is about ⅔ full.
2. Place the eye dropper upside down into the cylinder and allow it to come to rest in the water as shown in Figure O-11.

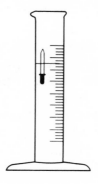

Figure O-11. *Technique for measuring comparative densities of different liquids using eye dropper and graduated cylinder.*

127

3. Record the reading on the cylinder at the very top of the inverted dropper.

4. Replace the tap water with sea water and repeat the procedure. Be sure that you use the *exact same quantity* of each liquid tested. In this manner, you can test and compare the densities of a variety of liquids such as those listed.

5. After measuring and comparing the densities of different liquids, test the change in density between cold and hot tap water. First measure the density of water cooled with ice cubes or in a refrigerator; then measure the density of water heated by a Bunsen burner.

ANALYZING RESULTS

This experiment demonstrates that sea water is denser than tap water and cold water is denser than hot water. Sea water, as demonstrated in the previous experiment, "What Makes Sea Water Heavy?", is heavier than tap water because it contains a greater amount of dissolved mineral matter. Thus it requires less sea water to equal the weight of the eye dropper, and the dropper floats higher in sea water than in tap water.

The density relationship of sea water to tap water is important in studies of sea water encroachment into underground fresh water storage supplies. It is also important when the flow of river water is studied as it enters the oceans. However, river water commonly contains a large percentage of fine rock and mineral material which it is transporting downstream. Much of the finer material may be carried suspended in the water. This will, of course, increase the density of the river water such that it may exceed that of sea water—as demonstrated in the following experiment.

The relationship of density contrasts between hot and cold water is also important as it relates to the convection in bodies of water (lakes or oceans). Such convection has a direct effect on the distribution of food for marine animal life. It, therefore, plays a major role in the actual distribution of the marine life itself. The relationship of density to temperature in sea water—and other fluids—will be further investigated in Oceanography Experiments 11 and 12.

INVESTIGATING FURTHER

Examine the relationship of density of water to its sediment content.

Add soil to a sample of tap water and mix thoroughly; then measure its density using an eye dropper as before. Compare this result to the densities of tap water and sea water. Do you see any relation to the currents which flow along the floor of the continental shelf outward from river mouths toward the continental slope? Some of the currents have been traced down the continental slope to the deep-ocean floor.

Temperature, amount of material in solution (salinity of sea water), and the amount of suspended rock and mineral matter contained all relate to the density of sea water. It is, therefore, to be expected that the ocean waters display density differences from place to place—and they do. How do these features relate to major ocean currents—such as the Gulf Stream, the North Atlantic Drift, and the North Pacific Drift? (Prevailing winds are a major factor in creating surface ocean currents; in contrast, deeper ocean currents largely develop in response to water density contrasts and to the earth's rotation.)

10

Density Currents

When a hurricane approaches the shore, high waves and strong winds are readily apparent. The beach front may be drastically modified with extensive erosional changes. But what happens to material removed from the beach or disturbed on the shallow sea floor? Submarine cables are sometimes broken on the continental slope, and on occasion successive cable breaks have occurred in a short time. Some of the cable breaks took place during the time when hurricanes have moved from the oceans onto the continents. Perhaps the breaking of submarine cables is closely related to the hurricanes and sediment they remove from the beach and shallow sea floor.

BACKGROUND INFORMATION

Currents formed as a result of contrasting fluid densities are known as *density* currents. As demonstrated in the previous experiment, the density difference may result from temperature contrasts or from varying amounts of material contained in solution or suspension. Of special importance to marine geology are currents which contain abundant sediment.

130

Known as *turbidity currents,* they are a major process in marine deposition.

Turbidity currents are generally initiated in response to sediment slump. This commonly occurs after a large mass of sediment is rapidly deposited, for example as a delta where a river enters a still body of water (lake or ocean). The delta of the Mississippi River in the Gulf of Mexico is an excellent example. Through continued sedimentation, the sediment mass becomes unstable and a slump is triggered—either by additional sedimentation, an earthquake, or perhaps a hurricane. Sediment of the slumping mass mixes with water and thus forms a turbidity current whose density exceeds that of the surrounding water.

Turbidity currents commonly form in this manner on the continental shelf. Because their density exceeds that of the surrounding sea water, they hug the shelf bottom as they flow seaward. These currents are then channeled down the steeper continental slope, flowing in submarine canyons which are carved out of the slope. The flows come to rest on the abyssal plains of the ocean floor.

Although they may be set in motion by sudden impulses such as an earthquake or hurricane waves, most turbidity currents are fed sediment from a relatively constant source for up to several hours. The currents commonly attain thicknesses of 100 m or more, and may be as much as 5 km long. Speeds of flows average up to 10 km/hr and, as they accelerate down the continental slope, may reach speeds of a few tens of m/sec.

Many submarine cable breaks have been attributed to turbidity currents. One such example which is well documented is off the Grand Banks along the southeast coast of Newfoundland. There, an earthquake in 1929 caused a slump on the continental slope and a turbidity current resulted. In an orderly sequence, the current broke submarine telegraph cables stretching along the abyssal plain of the nearby ocean floor. From this and other well-documented turbidity current flows, their importance as sediment-transporting features is now established.

MATERIAL

- Plastic shoebox (or small aquarium)
- Large jar or beaker
- Dirt
- Sea water
- Food coloring

131

STUDENT PROCEDURE

1. Place the shoebox (or aquarium) on a table and prop up one end about 5 cm.
2. Pour tap water into the shoebox until it is about ⅔ full.
3. Add several drops of food coloring to a container of sea water and mix.
4. Slowly pour the colored sea water into the shoebox of tap water (see Figure O-12). Pour so that the colored sea water flows down the inside of the box (at the high end). Watch how it flows along the bottom of the box, moving as a density current. Be sure to view from the side of the box so as to observe that the lighter tap water is above the heavier sea water.
5. To produce a turbidity current, place the shoebox on the table as before. This time add sea water until the box is about ⅔ full. This will represent a still body of salty water like the ocean.
6. Now add several handfuls of fine dirt to a container of tap water and mix thoroughly. This will represent a muddy river entering the ocean.
7. Slowly pour the muddy water into the shoebox, allowing it to flow down the inside of the high end of the box as before (see Figure O-12). Then you can observe a turbidity current flowing under sea water.

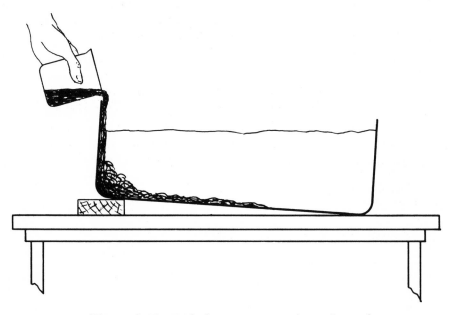

Figure O-12. *Turbidity current formed with muddy water poured into a tank of sea water.*

ANALYZING RESULTS

The two experiments produced currents which formed as a result of density contrasts as heavier water flowed into a body of lighter water. In the first experiment, sea water (heavier because of the content of dissolved matter) was poured into lighter tap water. In the second, muddy water (heavier because of the content of fine dirt suspended in it) was poured into sea water.

Both experiments can be directly related to conditions in nature. In zones of warm, dry climate, increased salinity of ocean waters may result from rapid evaporation. These zones of increased water salinity may be in belts within the oceans or they may occur in areas of restricted water circulation (such as in bays and lagoons). The volume of water with higher salinity is, therefore, heavier than the surrounding water, and can cause density currents. The analogy of muddy water flowing in the seas has previously been described. It is a major method of sediment transport from the continental shelf down the continental slope and onto the abyssal plains of the ocean floor.

INVESTIGATING FURTHER

We have already seen that temperature can be a controlling factor on fluid density. In areas such as the polar regions, the ocean waters are considerably colder than they are nearer the equator. Colder (heavier) polar waters produce density currents as they flow toward warmer (lighter) tropical waters. To demonstrate this type of density current, chill salt water (with ice cubes or in a refrigerator) and pour into a shoebox of sea water which has been warmed by heating with a Bunsen burner. Add food coloring to the cold sea water so that you can see this chilled water as it flows under the warmer sea water.

Density currents flowing laterally are only one type of effect developed in response to temperature differences in sea water. Another is the vertical flow of water in a "convection cell" as is demonstrated in the next experiment.

133

11

Convection Cells

Growth of the mid-oceanic ridges and spreading of the sea floor result from movements below the earth's crust. Convection in a cell-like pattern within the mantle has been used by some geologists to explain the nature of these movements. Lateral flow of the upwelling currents within the mantle may result in dragging of the relatively thin oceanic crustal layers. These layers are rifted in the mid-oceans, spread laterally toward the ocean sides, and piled up and rammed under the edges of the continents. If convection within the earth's mantle is indeed a reality, how might its origin be explained?

BACKGROUND INFORMATION

Temperature alone can affect the density of water as was demonstrated in the previous experiment. These density differences can in turn cause convection—thus the overturning of water in the oceans as cold water from the polar regions mixes with warmer water at lower latitudes. The colder, denser water sinks below the warmer, lighter water and then moves toward the equator. When it becomes warmed, the water rises.

134

Material within the mantle of the earth is believed to behave in a similar fashion. Although considered to be a solid, rock in the mantle may flow because of the high temperature and pressure there. Flowage quite likely takes place in response to the earth's *geothermal gradient* (the gradual increase in temperature from the earth's surface to its center). The geothermal gradient of the crust (and perhaps upper mantle) is estimated at 1°C for every 30 m. It is theorized that convection may be occurring from bottom to top of the mantle in numerous "convection cells." As hotter material reaches the upper mantle, heat is lost through the crust. Thus cooled, the material convects back downward (Figure O-13).

Recent measurements of heat flow through the earth's crust support this theory. The sensitive heat-detecting devices now being used have outlined zones of anomalously high heat-flow values and interestingly, the mid-oceanic ridges are along such zones. Concentration of heat along the ridges, as zones of convectional upwelling, would explain the amounts of recent volcanism concentrated there. It could also explain the apparent lateral flow of crustal material away from the ridge crests, if indeed convection cells moved laterally at depth and the crustal layers merely reflect this movement by being "dragged along."

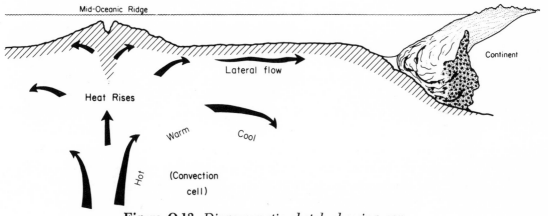

Figure O-13. *Diagrammatic sketch showing convection cell theory with material upwelling under mid-oceanic ridge and flowing downward under continental margin. Ocean floor slides away from ridge toward continent.*

MATERIAL

- Large beaker

- Bunsen burner

- Food coloring

- Sawdust or fine wood chips

STUDENT PROCEDURE

1. Place the large beaker (about ¾ full of water) on a ring stand and heat slowly with a Bunsen burner. Have the Bunsen burner located so that the flame is on one side of the bottom of the beaker, not centered.
2. When the water is heated so that boiling just begins, add a few drops of food coloring to the water. Watch the convection pattern which the colored water outlines.
3. Now add a handful of fine dirt to the beaker and mix thoroughly.
4. Carefully spread a thin layer of sawdust (or fine wood chips) over the top of the water.
5. Heat *very slowly* and observe the flow of the fine dirt in a convection pattern. What happens to the sawdust as boiling takes place?

ANALYZING RESULTS

This experiment demonstrates convection as a function of temperature. Heated water rises, cools, and moves downward as more heated water flows upward to replace it. Pattern of the flow can be readily followed by the movement of the colored water. A similar circular path is theorized for convection cells within the earth's mantle. The sideways movement of the sawdust and its piling up on the edge is characteristic of the movement of the earth's crust. As oceanic crustal material reaches the edges of the ocean basin, it is suggested that it is dragged downward under the continental edge (Figure O-13). New oceanic floor is in turn formed along the mid-oceanic ridges where volcanic rock is extruded.

INVESTIGATING FURTHER

A similar pattern of flow exists in the atmosphere. Differential heating of air takes place as a result of variations in solar energy absorption and reflection by different materials at the earth's surface (bodies of water, barren ground and bedrock, densely vegetated areas, etc.). For example, during sunny days the ground heats up more rapidly than do bodies of water. Cooler (denser) air over water flows landward as a gentle sea breeze to replace warmer (lighter) air on land. This warmer air in turn rises, cools, and moves seaward at higher altitudes. Indeed, the worldwide pattern of winds and pressure belts is a function of differential heating and consequent circulation.

Construct a wind-circulation cell to investigate this process. In an aquarium or similarly shaped tank with glass sides, place a large pan of water on one side (large enough to cover about ½ of the bottom of the aquarium). Cover the bottom of the other half with a layer of dirt, approximately 1 in thick. Place a lamp over the aquarium, directed at the side with the dirt bottom. Allow heating for several minutes, then place a small smudge pot releasing smoke in the center of the aquarium and move the lamp so that it continues to shine through the side of the aquarium directed toward the dirt. Place a lid over the aquarium. Watch the pattern taken by the smoke. It will move along the bottom toward the side covered with dirt, and then flow upward. Near the top it will flow across the aquarium and then downward above the water completing a circular pattern.

To prove that temperatures differ with the aquarium, place one thermometer in the dirt and one in the water. Note that the thermometer in the dirt records a higher temperature. The dirt absorbed more of the lamp's heat energy, and as it became warmer, the air above it was warmed.

Convection in response to temperature thus exists in the atmosphere, the oceans, and most probably in the lithosphere (within the mantle of the earth). The process is basically the same, although the medium changes in density and rate of flow.

12

Water Circulation in
Lakes and Oceans

MOTIVATORS

In contrast to almost every other substance, water can be more dense as a liquid than as a solid (ice). Can you imagine how different our world would be if ice was heavier than water? It is true that we would have no worry about water pipes and car radiators bursting when the water in them froze, but what about lakes and streams? They would freeze from the bottom upward into solid blocks of ice and aquatic life would not be able to survive. What other strange conditions can you imagine if ice sunk in water?

BACKGROUND INFORMATION

Most substances follow the general rule that their density increases as the temperature is reduced. Water follows this pattern as it cools to about 4°C. However, below 4°C water gradually expands until it reaches 0°C (freezing); it then expands greatly, adding about $\frac{1}{11}$ to its volume. It is this property that allows water in lakes to be zoned—commonly with a warmer surface layer above water which may be quite cold.

In the oceans, salinity (content of salts) is also important in controlling water density. Very cold water, made more saline by the freezing out of ice in Antarctica or Greenland is dense enough to flow along the ocean floor. In this manner, cold bottom-seeking currents which flow toward the equator from the polar regions are produced. The densest ocean currents result from a combination of low temperature and high salinity.

Because of the density contrast, there may be very little circulation across the boundary between the lighter surface layer in a lake or ocean and the denser water beneath. With increased warming, as in early summer in lakes, the density contrast between the layers becomes more and more distinct. They may, in fact, remain as separate layers until the following winter when once again the surface is cooled. When the point is reached that the surface waters are colder than the water beneath, convective overturning will occur.

The importance of density layering in a body of water is readily apparent when pollutants are dumped into it. Polluted water flowing into a lake or ocean may be confined to the surface zone, or may immediately sink into deeper water. Each type of pollution may cause major changes in the system's ecological balance.

In this experiment, density layering is produced to simulate these features in a body of water.

MATERIAL

- Small aquarium or glass-sided tank
- Ice
- Small portable hair dryer
- 5 Centigrade thermometers
- Food coloring

STUDENT PROCEDURE

1. Tape five thermometers along the inside of one side of the aquarium. Place them so they are at different distances from the top, and will therefore measure temperatures at different depths within the aquarium as shown in Figure O-14, page 141. Be sure to place the thermometers so they can be conveniently read looking through the glass and have one thermometer close to the water surface.

2. Add cold water to the aquarium until it is about ¾ full.

3. Add ice and stir occasionally, waiting for the water to be cooled to 4°C. When the water has reached this temperature, remove all the ice.

4. Turn on the hair dryer and allow the warm air to gently blow over the water from one end, aiming the hair dryer to "bounce" the warm air off the water surface at an angle of about 30° from the horizontal.

5. Record the temperature readings of the thermometers at five-minute intervals beginning with T_1 as the initial reading when the ice was removed (see Table O-5). Continue this procedure for at least 40 minutes.

6. When the temperature of the uppermost thermometer has reached about 8°C, add several drops of food coloring along the top of the water. Note how the dyed water remains in the upper, warmer zone.

7. Continue to warm the surface water, but gently move the dryer back and forth, across the surface of the water. This gentle motion will set up an oscillation in the water known as a *seiche* (say-sh). The water can also be set into a rocking motion by carefully raising and lowering one end of the aquarium. In both cases, the water motion will not affect the thermal boundary which has been created.

8. Take final temperature readings and note that the dyed water remains unmixed with the colder water beneath, even though the oscillation exists.

9. Graph the temperature readings of each thermometer.

Table O-5. *Record of temperature readings during development of density contrast by surface heating.*

Initial Temp.	Temperature readings at five-minute intervals						Final Temp.
T_1	T_2	T_3	T_4	T_5	T_6	T_7, etc.	T_F

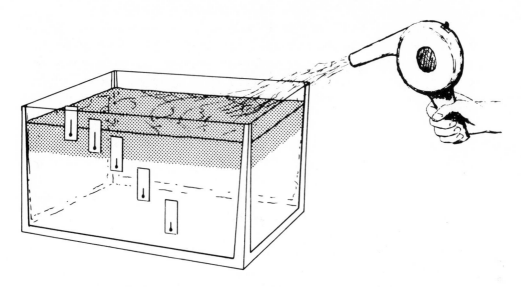

Figure O-14. *Technique for producing tempera-ture and density contrasts comparable to that found in lakes and the oceans.*

ANALYZING RESULTS

Temperature contrasts quickly develop between near-surface water and denser water beneath. Once density layering is established, circulation across the boundary is restricted. This is clearly observed after the food coloring is added to the warmer surface layer. Even with a seiche developed, the density (and temperature) contrast persists.

This can be a formidable barrier in the natural circulation of a lake or ocean. If the oxygen in the water below the boundary is depleted, the lack of circulation with surface waters prevents a replenishment of the supply. Thus the life forms which inhabit this lower environment will die for lack of oxygen, even though the upper zone may contain abundant oxygen.

This also accentuates the dangers of pollution. Sewage dumped into a lake may quickly concentrate in the lower waters, and if there is no chance for aeration, make these waters too toxic for life.

This experiment illustrates that once the boundary is established, further surface heating only accentuates the temperature (and therefore density) contrast. Have you ever gone swimming in a lake and felt the cold water at your feet when the surface water was comfortably warm?

Creating a seiche by moving the hair dryer over the water surface is, of course, analogous to the oscillation developed by prevailing winds. In contrast, the waves generated by raising and lowering one end of the tank may be compared to tidal waves produced by an earthquake. In this latter situation, the container itself moves rather than just the air above the water surface.

INVESTIGATING FURTHER

An analogy to water pollution can be made by adding muddy water to the aquarium. Note how this muddy water promptly sinks into the lower part of the aquarium. Can you imagine how toxic this water may become —especially when it is prevented from mixing with the atmosphere because of the density barrier to the upper layer?

After completing the experiment as previously outlined, gently place ice into the water. Watch the temperature on the thermometers in the aquarium. What happens to the colored water in the upper layer as it cools below that of the water beneath? Mixing will occur at this temperature because the near surface water is heavier than the water beneath. (This is analogous to convective overturning which occurs in some lakes in early winter when surface waters are cooled.)

Section Three

METEOROLOGY

1

Measuring Dew Point
and Humidity

W hat do most people look for when they read the weather report in the newspaper? Many of them simply want to know whether it will rain or not. Students can begin a study of rain-producing factors by looking at the formation of water droplets on a cold surface. Ask students to give examples of water condensing on cold surfaces. Where does this water come from? Does water collect on all cold surfaces, or just on some favored few? What can we do to find the answer to these questions?

BACKGROUND INFORMATION

This experiment requires a minimum of background information on the part of the student. Nevertheless, it is an admirable vehicle for raising questions about the many factors which influence the formation of rain, and a good introduction to the study of meteorology. The method for determining dew point and the equipment involved are both very simple. A quick call to the weather bureau will let students check their results.

The condensation of water vapor in the air into a water droplet requires some structure to serve as a center on which the water can condense—a speck of dust or something solid, known as a condensation nucleus. In this experiment the tin can serves as a giant condensation nucleus. The cooling is accomplished with crushed ice. (In Meteorology Experiment 7 we will investigate more natural ways in which the air might be cooled.)

Condensation by itself does not usually lead to precipitation. The tiny condensation droplets must become larger before they fall. One way this can come about is by collision of the droplets. If students watch the can long enough, they may see the droplets grow large enough to collide and combine into a drop big enough to run down the side of the can. A second way that condensation droplets can merge is by diffusing onto nearby ice crystals.

MATERIAL

- Tin can (with shiny surface)
- Glass jar
- Thermometer
- Large cardboard box
- Spray gun or bottle (e.g., the spray bottle of a liquid cleaner)
- Ice (crushed, or small cubes)

STUDENT PROCEDURE

How cold must a surface be before water will begin to collect on it? Students can find the temperature at which water *first* begins to form on the outside of the tin can by following these steps:

1. Pour water into the can until it is about ⅓ full.
2. Gradually add ice to the water in the can. Use the thermometer to stir the ice-water mixture so that you can be sure that the tin can is being cooled just as much as the water inside it.
3. As the metal can gets colder and colder, watch carefully for the first appearance of moisture on the outside. When it appears, read the temperature registered by the thermometer.
4. Check to see that you really did find the very *first* temperature at

which moisture formed on the can by pouring out the ice water, drying the can, and repeating steps 1-3. Keep repeating the experiment until you can decide on a definite temperature.

The temperature at which moisture first forms on a cold surface is called the *dew point temperature*. Does the value of the dew point temperature depend upon the kind of surface which is cooled? Try the experiment again with a glass jar. How can you be sure that the outside surface of the jar is at the same temperature as the ice water inside?

What does the dew point temperature have to do with the amount

Figure M-1. *Humidity can be increased by spraying water into a parcel of air surrounded by a large cardboard box.*

147

of moisture in the air? Students should not consider the experiment completed until they have had a chance to gather some evidence which will help them answer this question. What is really necessary in resolving the question is to change the moisture in the air. This can be done on a large scale by going into a bathroom where the shower has been running. However, the moisture can be changed in your own classroom in the following way:

1. Place a large cardboard box on its side.
2. Fill a spray gun, bottle, or even a water pistol with water and spray it directly into the air in the box. It won't matter if the sides of the box get wet; the only purpose of the box is to keep the wet air from blowing away.
3. Now put the tin can inside the box and find the temperature at which moisture *first* collects on the outside as shown in Figure M-1 (previous page).

ANALYZING RESULTS

What happens to the dew point temperature as the amount of water in the air increases? By now, students should be able to give a very general answer to that question. They can get a more detailed answer by collecting enough data to plot a graph. The amount of water in the air can be estimated by counting the number of squirts they pumped into the air with the spray gun. Then the dew point could be measured for air that contained 5, 10, 20, and 30 squirts of moisture. (If the entire class is doing this experiment, time can be saved by assigning one "squirt number" to each laboratory group.)

INVESTIGATING FURTHER

Because metal is a good conductor of heat (and cold), a metal can makes a good surface to cool in determining dew point. Moisture will condense just as well on other surfaces, but the temperature of those surfaces cannot be measured as easily. The teacher might want to begin a post-laboratory discussion by observing that the air isn't full of tin cans for collecting water. Does moisture have to have something on which to collect before it can rain, or was this just something artificial in our experiment? Suppose you had a sealed glass jar full of "wet air" and cooled it in a refrigerator. What would happen? Try cooling a sealed jar which has had some smoke blown into it.

Moisture in the air is called *humidity*. Relative humidity is a comparison of the amount of water in the air to the maximum amount which the air could hold under the same conditions. The most important variable involved is temperature.

Table M-1 relates the temperature of the air to the saturated vapor pressure (a kind of measure of the maximum amount of water vapor which the air can hold at that temperature).

Table M-1. *Saturation vapor pressure versus temperature.*

Temperature (°C)	−10	−9	−8	−7	−6	−5	−4	−3	
Saturation Vapor Pressure (millibars)	2.6	2.9	3.2	3.5	3.8	4.0	4.3	4.7	
Temperature (°C)	−2	−1	0 1	2	3	4	5	6	7
Saturation Vapor Pressure (millibars)	5.1	5.6	6.1	6.5	7.0	7.5	8.1	8.7	9.3
Temperature (°C)	8	9	10	11	12	13	14	15	
Saturation Vapor Pressure (millibars)	10.7	11.4	12.2	13.0	13.9	14.8	15.8	16.9	
Temperature (°C)	16	17	18	19	20	21	22	23	
Saturation Vapor Pressure (millibars)	18.0	19.2	20.5	21.8	23.2	24.6	26.1	27.8	
Temperature (°C)	24	25	26	27	28	29	30		
Saturation Vapor Pressure (millibars)	29.6	31.4	33.4	35.5	37.6	39.7	41.9		

The relative humidity is the ratio of the saturation vapor pressure at dew point temperature to the saturation vapor pressure at air temperature (converted to per cent).

The heights of clouds can also be estimated from knowledge of the dew point temperature, since the temperature of the air at the minimum cloud level must be the same as the dew point temperature. On the average, the temperature of the air drops 5.5°F for every 1000-ft rise (in comparison to the temperature of the air at ground level). The dew point temperature also drops with a change in altitude because the amount of water vapor which the air can hold depends upon the air pressure. The dew point falls at the rate of 0.9°F for every 1000-ft rise (in comparison to the dew point of the air at ground level). The first height at which clouds can form is then the height at which the air and dew point temperatures are equal. This is given by

$$T_g - 5.5(h) = D_g - 0.9(h),$$

where T_g is the ground temperature of the air, D_g is the ground dew point temperature, and h is the height in thousands of feet.

The equation can be simplified to

$$h = \frac{T_g - D_g}{4.6} \quad .$$

(Remember that this gives the height in thousands of feet. For example, if h = 5, the lowest cloud height is 5000 ft.)

2

Evaporation

In the previous experiment, students increased the moisture in the air by spraying water into it with a spray gun. Obviously, there must be other ways to increase the humidity apart from direct spraying. In this experiment, students can investigate factors which affect the rate at which a water-saturated sponge gives its water to the air. The water loss will be seen by weighing the wet sponge and watching its weight loss.

BACKGROUND INFORMATION

Three basic factors affect the evaporation rate of water: (1) a difference in the energy available to supply the heat of evaporation; (2) the size of the surface area of the interface between the wet body and the air; and (3) the prevention of recondensation of the evaporated water. The last factor is affected by the humidity in the air and by air currents which carry the water vapor away from its source.

Students may identify several different factors in this experiment, but in general they can all be assigned to these three general classifications.

151

MATERIAL

- 2 similar sponges (approximately 7 x 12 x 1 cm)
- Graduated cylinder (or graduated medicine bottle)
- Light bulb (100 watts or greater)
- Light bulb socket and cord
- Meter stick
- String

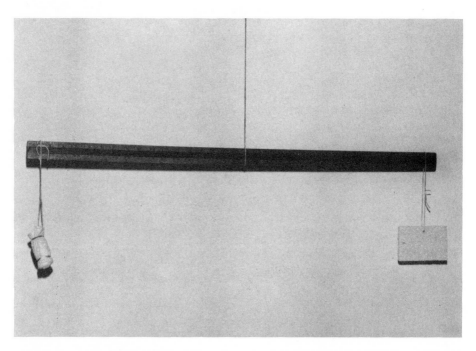

Figure M-2. *The rates of evaporation of water from two sponges with different surface areas can be compared using a meter stick balance.*

STUDENT PROCEDURE

We want to watch the water loss by two wet sponges under different conditions. As sponges lose water they become lighter, so we can use a simple meter stick balance to compare the rate at which they are losing water.

1. Tie a string around the center of the meter stick so that the stick remains level when it is supported by the string. Tie the other end of the string to some overhead support.
2. Prepare the sponges so that they can be hung from the meter stick. Punch a hole through one corner of each sponge, and thread a piece of string through each hole. Tie the ends of the string so that the resulting loop will slip over the ends of the meter stick as shown in Figure M-2.
3. Wet the sponges equally. This is best done by pouring equal amounts of water on each sponge. The amount of water can be measured in the graduated cylinder or medicine bottle. The exact amount of water needed depends upon the size of the sponges you are using. The sponge should be wet, but not dripping.
4. Hang the wet sponges on opposite ends of the stick, sliding them back and forth until the stick balances once again. Place a bright light bulb beneath one of the sponges. (Leave the other sponge alone.) Wait a few minutes. Which sponge loses water more rapidly?

You can wring the sponges so that they are as dry as you can get them, and then repeat the experiment. Try changing the following factors one at a time:

1. Fan the air around one sponge with a piece of paper.
2. Roll one of the sponges up like a jelly roll before you wet it. Tie it in this shape with two pieces of string, then pour on the water.
3. Place a box around one of the sponges. Increase the humidity of the air in the box by spraying water, as was done in the previous experiment. (Be sure not to get any additional water on the sponge.)

ANALYZING RESULTS

When water is lost directly to the air, we say that it has evaporated. Have the student list factors which were changed in this experiment which affected the rate of evaporation. Ask how one can be sure that the differences were due to the warming, fanning, surface area, etc. and not to some difference between the sponges themselves.

INVESTIGATING FURTHER

Does warming make a difference because the air around the sponge gets warm, or the water in the sponge gets warm, or the sponge itself gets warm? Let the students pour hot water on one sponge and an equal amount of cold water on the other and repeat the experimental procedure.

Students can also wrap a wet cloth around a bulb of a thermometer and in a few minutes compare its temperature reading to that of a plain thermometer. Suggest that this comparison can be used as a measure of humidity in the air, and ask students what evidence in this experiment would tend to support that suggestion.

3

Absorbing Warmth

MOTIVATORS

Humidity is one of the major factors in personal evaluation of weather. In the two previous experiments, students have seen how humidity is measured and how evaporation and humidity are related. Perhaps some of them noticed that the warmth from the light bulb made a difference in the rate of evaporation. Different objects on our earth are warmed by the sun in much the same way that the sponges were warmed by the light bulb. How do different materials compare in absorbing warmth? Does the color of the material make a difference in the absorption of warmth? Students can find the answers to these questions in the laboratory.

BACKGROUND INFORMATION

Almost all the energy that reaches the earth's surface and atmosphere comes from the sun. Differences in the absorption of this energy are a primary source of the imbalances which cause our constantly changing weather patterns. This experiment begins a sequence of four experiments which investigate the effects of transferring and absorbing heat energy.

155

Meteorology Experiments 3 through 5 use the words "warming" and "warmth" in referring to heat and temperature. Most students will not consciously make a distinction between heat and temperature when they begin this experiment. By the end of Experiment 5 they should realize that some distinction is necessary, and Experiment 6 is designed to make that distinction explicit.

The exact atomic mechanism which relates the absorption properties of different substances to their color is very complex and not well understood. The purpose of this experiment is simply to show that differences in absorption rates do exist. All of the gases, clouds, and dust in the atmosphere absorb about 19% of the incoming solar radiation. The cloud cover reflects about 25% back into space, and reflection from other sources such as gas molecules and the earth's surface account for another 9%. This leaves 47% of the incoming solar radiation to be absorbed by materials at the earth's surface. The different rates of warming of land areas and water areas is a major factor in determining climates and weather patterns.

MATERIAL

- 6 styrofoam drinking cups
- Large (200 to 300 watts) light bulb, light bulb socket, and cord
- Thermometer
- Food colors (red, green, and blue)
- Black dirt, and sand

STUDENT PROCEDURE

We want to see how several different objects warm up when heated for the same time by the same heat source.

1. Fill the styrofoam cups with about the same amounts of samples. Try plain water, green-colored water, blue (or ink)-colored water, red-colored water, black dirt, and sand. These samples will give us a good range of color variation as well as variations in materials.
2. Hang up the light bulb and arrange all the samples in a circle below it so that they are all equally distant from the light bulb (about 10 cm away) as shown in Figure M-3.
3. Note the time, and turn on the light bulb.

Figure M-3. *Samples of sand and water absorb heat at different rates even when placed at equal distances from a common heat source.*

4. After about 10 minutes measure the temperature of each of the samples. Since the top will be warmer than the bottom, it is important to measure the temperature at the same relative position each time. You can do this by putting the thermometer in the sample until the bulb is just covered in every case.
5. Wait another 10 minutes and then take another set of readings.
6. Continue this process every 10 minutes until you can be sure that you have found differences between your samples.

ANALYZING RESULTS

Have students record their observations in a table so that they can see a pattern emerging. They might want to make a graph of temperature versus time for each substance. Let students discuss the effect of color and material upon the rate of warming as shown by their data.

INVESTIGATING FURTHER

Students were instructed to use about the same amounts of all samples. Some of them may want to investigate what would happen if the amounts used were drastically different. Suppose they used a cup full of water and a cup with just enough water to cover the bulb of the thermometer. Which would have warmed more? Of course, the surface of one water sample would have been farther away from the light bulb than the other, but some might have been ingenious enough to make the differences equal by setting one of the cups on a block.

Ask students what would happen if several samples were placed at varying distances from the light bulb. Would the sample twice as far away have half the temperature change in a given time? Would the angle at which the light rays hit the surface of the sample affect the results? (The angle can be estimated from the length of the shadow of the styrofoam cup on the desk top.) Students may want to extend the investigation by varying these factors.

Finally, pose this situation: suppose one had a bathtub full of water at 50°C and a styrofoam cup full of water at 50°C. Are they both equally warm? Would it take the same time to warm both of them from room temperature up to 50°C? Is there a difference between warmth and warming?

4

Transmitting Warmth

In the previous experiment, students investigated the warming of several different substances and colors. The source of that warmth was an ordinary light bulb. But how did the warmth travel from the light bulb to the samples in the styrofoam cup?

Warmth can travel by touch or conduction. For example, if one touched a long nail to the light bulb he would soon find that the far end of the nail grows warm. In this case, the warmth travels down the nail by conduction. Yet, in the last experiment, nothing touched the light bulb and sample to form a path for conduction of the warmth. Students will immediately rule out this method of warming as an explanation of what happened in the preceding experiment.

What other choices do they have? The only thing between the light bulb and their samples was the air itself. Perhaps they should begin by looking at the motion of the air around a hot light bulb.

BACKGROUND INFORMATION

This experiment investigates two ways in which heat might travel

159

from the light bulb to the sample which is warmed. If your students left the sample sitting under the light bulb long enough in the previous experiment, they may have noticed a time when the temperature of the sample was emitting heat just as fast as it was absorbing heat from the light bulb. Therefore, this experiment could have been motivated by asking how the *sample* transmitted warmth.

Like the sample in the styrofoam cup, the earth cannot simply absorb solar radiation. If it did, its temperature would continue to rise rather rapidly as time went on. To maintain a thermal equilibrium, it must radiate the energy which it absorbs back into space. Thus the earth's surface also acts as an additional energy source to the cooler air above it. This source produces convection circulation in the atmosphere much like the convection students will study in this experiment.

MATERIAL

- Large (200 to 300 watts) light bulb
- Light bulb socket and cord
- Large cardboard box
- Clear plastic wrap
- Sheets of typing paper
- Lemon juice or onion juice
- Flat pocket mirror
- Magnifying glass or convex lens
- Matches or wooden splints

STUDENT PROCEDURE

Students can use the cardboard box to confine the air around the light bulb and shield it from stray wind currents. The following steps detail the construction of a box like the one shown in Figure M-4.

1. Put the box on its side so that what was the open top now faces you. Into the new top of the box cut two holes, as far apart as possible, and large enough so that the light bulb and socket can be lowered through either one into the box.

160

Figure M-4. *A cardboard box covered with plastic wrap can be used to study convection currents around a hot light bulb.*

2. Cover the open face of the box with clear plastic wrap, taping the covering to the edges to make it secure.
3. Suspend the light bulb and socket through one of the holes so that the bulb hangs about halfway between the top and bottom of the box.
4. Turn on the light bulb and wait a few minutes.
5. Now place a smoking object at the other hole and watch the movement of the smoke as it is carried by the air. (For the smoking object you can use matches or wooden splints which have been ignited and then blown out. Or you can soak some paper in a weak solution of potassium nitrate, let the paper dry, and then burn it to produce smoke.)

Have the students draw a sketch showing the movement of the air within the box. Does their sketch indicate that the air which was warmed by the light bulb would have carried that warmth to the samples used in the previous experiment? Remember that the samples were placed *below* the heat source.

Students can try something else. Many of them may have written secret messages using lemon juice or onion juice. They will remember that these juices dry colorless on the paper. A message written with such juices will be invisible until the paper is warmed. When heated, the lemon juice markings turn brown and the message can be read.

Students can use this same idea to get an indicator of warmth which will allow them to locate patterns of heat transmitted by radiation.

1. Spread lemon juice or onion juice over several pieces of typing paper and let them dry.
2. Place a piece of the dried indicator paper underneath the light bulb. The light bulb should be no more than 1 or 2 cm away from the paper. Wait for several minutes. The places which turn brown first indicate where the greatest warmth is.
3. Put a new piece of indicator paper underneath the light bulb. Above the bulb and a little to one side hold a pocket mirror so that extra light falls on one side of the paper. Wait for several minutes. Which side of the paper turns brown first?
4. Put a new piece of indicator paper underneath the light bulb. This time raise the light bulb so that it is about 15 cm from the paper. Hold a magnifying glass so that it is between the light bulb and the paper. Adjust the position of the glass so that extra light shines on one spot of the paper. Which part of the paper turns brown first?

ANALYZING RESULTS

Students have looked at the transmission of warmth by both convection and radiation. They should be able to decide that radiation was the method which was primarily responsible for the warming of the samples in the previous experiment.

Have students look at the brown spots on the indicator paper. What can they say about the amount of light striking the paper and the amount of heat warming the paper?

INVESTIGATING FURTHER

Perhaps many students will conclude that in the case of the light bulb the warmth is greatest where the light is brightest. This conclusion suggests that we might be able to measure the warmth given off by a light bulb without using a thermometer. We could get a measure of the warmth given off by using a light meter. As we shall see in the next experiment, repeating this experiment with a light meter gives surprising results.

5

Reflecting Warmth

In the preceding experiment, students saw that the greatest amounts of heat given off by a light bulb were apt to be concentrated in the places where the illumination of light was the greatest. In this experiment, they will use this idea to help compare amounts of radiated heat. To measure the amount of light given off by the light bulb, they can use a photographer's light meter. If one inspects a typical light meter, he will probably see that the scale is not marked off evenly, but instead is marked with the f-stop settings of a camera. To use this light meter, the student will first have to construct a new scale which gives readings of light intensity directly. Instructions for making such a scale are given in Appendix IV.

BACKGROUND INFORMATION

In the previous experiment, "Transmitting Warmth," students saw reasons for believing that radiated warmth could be reflected in the same way that light can be reflected. What does this mean where the warming of materials is concerned? If a sample reflects most of the warmth it

163

receives, then it shouldn't warm up so fast (since there would be little heat left for it to absorb). This means that the materials which have the biggest temperature rises (absorb the most warmth) should be the ones which reflect the smallest amounts of warmth.

Students now have reason to believe that the amount of warmth reflected by materials can be measured by the amount of light they reflect. This means that the materials which absorb the most warmth should reflect the least light. If this is so, one can then walk around outside and predict that the objects which look dullest to us (reflect the least light) will also be the hottest (absorb the most warmth).

Is this true? Can students think of examples which agree with this hypothesis? Fortunately, it is not hard for students to check out this hypothesis in the laboratory. All they need to do is repeat the previous experiment quickly; this time pointing the light meter at the surface of each sample in turn to see which reflects the most light. That sample should warm the least.

MATERIAL

- Large light bulb (200 to 300 watts)
- Light bulb socket and cord
- Light meter, calibrated according to the instructions in Appendix IV
- 6 styrofoam cups
- Food colors (red, green, and blue)
- Black dirt, and sand

STUDENT PROCEDURE

1. Have students look back at their data for Meteorology Experiment 3. List the materials used in that experiment in the order of their ability to absorb warmth, starting with "best absorbers" and ending with "worst absorbers."

2. Place a material sample under the light bulb. Point the light meter at the surface of the sample, and move it upward and back until the needle points approximately to the middle of the scale. (If the reading increases as you move the light meter away, you are picking up light directly from the light bulb instead of from the surface of the sample. You may have to cut off this light with a cardboard shield.)

3. Fix the light meter in this position. You may want to stack books underneath it, or you may want to simply hold it and have a partner help you with the rest of the experiment. In either case, the position of the light meter (both height and direction) must *not* be changed throughout the rest of the experiment.

4. Record the light meter reading for the light reflected from the surface of the first sample.

5. Replace the first sample with the second sample. Again obtain a value for the light reflected from the surface.

6. Repeat the procedure until you have obtained a value for the light reflected from the surface of each of the six samples.

ANALYZING RESULTS

Using the readings of the light meter, have students list the samples they tried, beginning with "worst reflector" and ending with "best reflector." Our hypothesis indicates that the order of this list should be the same as the order of the list which begins "best absorber" and ends with "worst absorber." Is the order the same?

INVESTIGATING FURTHER

Many other people who have done this experiment have found that their two lists do not have the same order. Do your students agree? It is probable that they will find that the sample of sand was both the best absorber and the best reflector.

If this is the case, students are now faced with a dilemma. The best absorbers do not seem to always be the worst reflectors. Will we have to reject the hypothesis (which was really based on a conservation of energy intuition)?

There is one possible way out of this dilemma. One can get out of it if he can show that the thermometer and the light meter do not measure the same things. We have used the light meter as an indirect measure of the warmth (heat) coming from the light bulb. We have used the thermometer to measure the warming (temperature change) of different substances. Could it be that warmth and warming are different? Could it be that heat and temperature are not the same thing? Ask students to suggest possible experiments which might answer these questions. The next experiment is designed to emphasize the distinction between heat and temperature.

165

6

Heat and Temperature

MOTIVATORS

The sun heats the land; as a result the temperature of the land rises. Are these events the same or different? In this experiment, students will investigate the differences between heat and temperature.

BACKGROUND INFORMATION

Although this is not a precise experiment, students should not only be able to sharpen the distinction they make between heat and temperature, but should also be able to estimate values of specific heat for different substances. The standard unit of heat (1 calorie) is defined as the amount of heat necessary to raise the temperature of 1 gram of water by 1° Celsius. For this experiment we shall use as the unit of heat the amount of heat given off by the student's Bunsen burner in 10 seconds. As long as a student uses the same Bunsen burner at the same gas-jet setting and at the same height above the substance to be heated, this timing will give a definite amount of heat. That amount of heat cannot be compared with the amount of heat taken as a unit by another student, or even with the amount of heat taken as a unit by the same student on a different day (when the total conditions are apt to be different). Therefore, the comparisons called for in this experiment must be made by a single student in a single laboratory situation.

The *specific heat* of a particular material is the fractional amount of heat necessary to change the temperature of that material when compared to the heat necessary to cause the same temperature change in the same amount of water. That is, if the specific heat of material "x" is 0.50, then it takes just half as much heat to cause a certain temperature change in material "x" as would be required to cause the same temperature change in the same amount of water. In our experiment, this means that it should take just half the time to heat material "x." This means that we can estimate the specific heat values of different substances very easily. We weigh out the same amounts of each substance, and time how long it takes to raise the temperature of each by, say, 10°C. We also find the time required to raise the temperature of the same weight of water by 10°C. The specific heat, c, for the sample is then

$$c = \frac{\text{time (sample)}}{\text{time (water)}}$$

(for the same weights and temperature changes).

While it is important for the teacher to understand the basic scheme by which we can determine specific heat values before the beginning of the experiment, it is not necessary that the student understand this until after he has done the experiment. The experiment simply asks him to compare times required to heat two different substances. He can make this comparison by either subtracting or dividing the two numbers. The teacher can lead a discussion of specific heat most easily *after* the student has specific values in front of him to compare.

MATERIAL

- Small beaker (200 ml)
- Ring stand and wire gauze
- Bunsen burner (an electric immersion heater may be substituted)
- Thermometer
- Balances
- Stopwatch (or wall clock with a sweep second hand)
- Cooking oil and water

STUDENT PROCEDURE

We want to compare the amounts of heat necessary to cause equal temperature changes in materials which differ both in type and amount. As a unit of heat we shall use the amount of heat given off by a Bunsen burner in 10 seconds. To be sure that this is the same amount of heat every 10 seconds you must be careful *not* to change the gas supply nor the distance from the burner to the material to be heated once the experiment has been started.

Begin by investigating what happens to the amount of heat necessary to produce a 10°C temperature change when the amount of material which is heated is varied.

1. Weigh 50 g of water in a beaker. (Be sure to subtract the weight of the beaker from the total weight on the scale. You could save some work by remembering that 1 ml of water weighs 1 g.)

2. Light and adjust the Bunsen burner. Place a wire gauze on a ring stand. Place the Bunsen burner under the ring stand, adjusting the height so that the flame just reaches the wire gauze. *Do not change this height or the Bunsen burner flame once you have started the next step in the procedure.*

3. Place the beaker on the wire gauze and note the time and initial temperature of the water. Gently stir the water in the beaker with the thermometer.

4. When the temperature of the water is 10°C higher than the initial temperature, remove the beaker of water and read the time. Record the temperature change, the amount of water, the elapsed time, and the number of heat units supplied in a manner as shown in Table M-2. (Remember that 10 seconds = 1 heat unit.)

5. Rinse the beaker with cool water to bring it back to room temperature. Measure 100 g of water into the beaker and again heat it until the water temperature has risen 10°C, noting the beginning and ending times. When this is done record the temperature change, the amount of water, the elapsed time, and the number of heat units supplied for this second sample.

6. Again rinse the beaker with cool water. Find the time necessary to raise the temperature of a sample of 150 g of water by 10°C. Once again record the temperature change, the amount of water, the elapsed time, and the number of heat units supplied.

What relationship exists between the amount of temperature change and the number of heat units supplied? Can you predict (or estimate) how many heat units would be required to raise the temperature of 200 g of water by 10°C? Check your prediction if you think it is necessary.

Now let's see what happens to the heat units when we vary the kind of material that is heated.

1. Find the number of heat units necessary to raise the temperature of 100 g of water by 20°C. (Follow the same procedure that you did before. Record the temperature change, the kind of sample, the amount of the sample, the elapsed time, and the number of heat units supplied.)
2. Cool the beaker with water, and dry it. Weigh 100 g of cooking oil in the beaker. (Don't forget to subtract the weight of the beaker.)
3. Find the number of heat units necessary to raise the temperature of 100 g of cooking oil by 20°C. Record the temperature change, the kind of sample, the amount of the sample, the elapsed time, and the number of heat units supplied.
4. Pour the cooking oil into a waste container, and wash the beaker with soap and water. Dry the beaker and weigh 100 g of sand in it.
5. Find the number of heat units necessary to raise the temperature of 100 g of sand by 20°C. You can assume that the initial temperature of the sand is the same as the temperature of the room. You will need to stir the sand rapidly while it is heating. Remove the beaker of sand from the heat just before you reach the final temperature, since the temperature registered by the thermometer may continue to rise as the bulb touches different grains of sand. (It may be necessary to repeat this trial if the temperature goes too high.) Record the temperature change, the kind of sample, the amount of the sample, the elapsed time, and the number of heat units supplied.

ANALYZING RESULTS

This experiment provides good practice in organizing data so that it can be easily compared. You may wish to suggest the following table as an organizational aid.

Table M-2. *Data table for recording heat units and time required for temperature rises.*

Substance	Amount	Time for 10° Rise	Heat Units
water	50g		
water	100g		
water	150g		
Substance	**Amount**	**Time for 20° Rise**	**Heat Units**
water	100g		
cooking oil	100g		
sand	100g		

169

Ask students how the amounts of heat given to the cooking oil and the sand compare to the amount of heat given to the water. They can make a comparison for sand by dividing the number of heat units used to heat the sand by the number of heat units used to heat the water. The value of this fraction is called the specific heat of sand.

To find the specific heat of cooking oil, divide the number of heat units used to heat the cooking oil by the number of heat units used to heat the water.

What would the specific heat of water be?

INVESTIGATING FURTHER

A chemistry or physics textbook will contain a table of specific heat values for many different materials. Have students compare these values with the specific heat of water (which is 1.00 by definition).

Ask students whether a substance with a small specific heat value warms up rapidly or slowly. Since a small specific heat value means the material requires little heat to undergo a definite temperature change, the material would warm up rapidly. If the heat source is taken away, the material will also cool off rapidly in comparison to water. This is why the air around large bodies of water is cooler in the daytime and warmer at night. The vast areas of water which cover most of this planet tend to act as a "brake" on the warming-cooling cycle from day to night. For the same reason, the desert experiences extreme temperature changes from day to night because it is composed of materials with low specific heat values. The distribution of land and water areas thus affects the heating of the atmosphere, modifying the convection-cell patterns in the air. The rotation of the earth further modifies these patterns.

7

From Heat to Pressure

In the previous experiment, students heated liquids and solids to determine their specific heat values. These values determine the relative sizes of temperature changes when earth materials absorb radiation from the sun. Since the temperature of the soil does not continue to rise indefinitely, it must eventually radiate away this incoming warmth. This secondary radiation is an important source of heat for warming the atmosphere. What happens to the air when it is warmed? One answer to this question is that convection cell patterns are established (See Meteorology Experiment 4, "Transmitting Warmth.") In this experiment, students investigate changes in pressure when the air is warmed.

BACKGROUND INFORMATION

The gas laws of chemistry and physics (PV=kT) do not hold as expected in the atmosphere. These gas laws are applicable when the gas is confined in a container. Thus while heating gas in a container *increases* the pressure, warming of the atmosphere *decreases* the pressure since the gas is free to expand (decreasing its density accordingly).

MATERIAL

- Disposable (plastic) medical injection syringe (available from pharmacies)
- Beaker
- Ring stand and wire gauze
- Bunsen burner
- Test tube
- Balloon
- Tongs
- Pliers

STUDENT PROCEDURE

What happens to the air pressure when the air is heated? To find out, try the following test:

1. Fit a balloon over the mouth of a test tube. If the neck of the balloon is small enough, no additional fastener will be required.

2. Light and adjust the Bunsen burner. Using a pair of tongs, hold the lower end of the test tube in the flame. Be sure to hold the test tube so that the balloon is away from the flame.

What happens to the air in the test tube as it is heated? Consider the shape of the balloon before and after the air was heated. Is the pressure of the air on the balloon more or less after the air is heated?

Now try this experiment another way. Atmospheric air pressure is the weight of all the air above a unit area. Suppose we take the area of a single square floor tile as a unit area. Imagine a box with this floor tile as a bottom and vertical sides extending upward to the "top of the atmosphere." The air pressure at the floor level is the weight of all the air contained in this imaginary box.

The density of air is the weight per unit volume. To find the total weight of air in our imaginary box we should multiply the density of air by the number of unit volumes which would fill the box. (In our imaginary case, a unit volume would be a cube with exactly one floor tile for each face.) Every box we construct this way will contain the same number of unit volumes. Whether the air pressure will vary from box to box

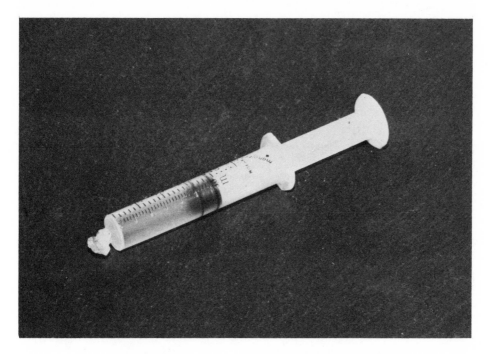

Figure M-5. *When a plastic syringe is heated the
air trapped inside expands, changing its density.*

depends, then, upon whether the density of air can vary. Does this density
vary with temperature? You can find out in the following way:

1. Take a plastic medical injection syringe. Push the plunger about halfway in. Heat the other end with a match (or low Bunsen burner flame)
   until the plastic begins to melt. When this happens, pinch the end
   shut with a pair of pliers, as shown in Figure M-5.
2. Record the volume of air which is trapped between the plunger and
   the sealed end (1cc $=$ 1 cm³).
3. Find the weight of the air trapped inside the syringe. The density
   of air in your classroom is about 0.0014 g/cm.³ The weight of the
   air in the syringe is equal to this density multiplied by the volume
   determined in step 2.
4. Boil water in a beaker. Drop the syringe in the boiling water and
   leave it for at least two minutes.
5. Take the syringe out of the beaker with the tongs. Read the volume
   of the heated air trapped inside.
6. Calculate the density of the heated air by dividing the weight of the
   air (in step 3) by the volume of the air (in step 5).

173

ANALYZING RESULTS

Ask students if the density of the air increased or decreased when it was heated. Have them think back to the imaginary box. If this box was filled with heated air would the pressure be greater or smaller?

Have students record their observations by writing a paragraph describing what happened. They should also show the calculation of the weight of the air trapped in the syringe and of the density of the heated air.

INVESTIGATING FURTHER

The syringe could also be cooled in a freezer as a demonstration of the continuation of this effect when air is cooled instead of heated.

Ask students to compare the results obtained in the first part of this experiment (balloon) with the results obtained in the last part (syringe). Do these results agree? See if your students can formulate an explanation of the discrepancy.

The heated air could not expand in the balloon until its pressure increased enough to stretch the rubber. But if the plunger of the syringe was free to move, no such pressure increase needed to occur before the air could expand in the syringe. In the atmosphere, air is not confined and is therefore free to expand as it is heated. Since a given volume of this heated air weighs less, it can exert less pressure upon a surface.

In this experiment, students changed air pressure, but never measured it directly. Have students discuss how air pressure is measured. Ask how the atmospheric pressure over the equator would compare with the air pressure over the poles.

Think back to the imaginary box. Suppose the air was distributed uniformly throughout the box. What would the pressure be on a floor tile that was inserted halfway up the box, compared to the pressure on a floor tile at the bottom of the box? If the pressure is caused by the weight of air *above* a unit area (the floor tile), what would students predict about the value of atmospheric pressure at higher and higher altitudes? (Because of the earth's gravity, the air is not distributed uniformly through the imaginary box. Most of it is concentrated at the bottom with the concentration becoming less and less as one ascends. The effect of this is merely to reduce the atmospheric pressure more rapidly with an increase in altitude than would normally be expected.)

8

From Pressure to Cold

Have students take the syringe of trapped air used in the previous experiment and push in on the plunger. Ask them to compare the force necessary to compress the air from 2.0 cc to 1.9 cc with the force necessary to compress the air from 1.0 cc to 0.9 cc.

Have the students pull out the plunger and fill the sealed syringe with water. Let them try to insert the plunger back into the water-filled syringe. Ask them if they can compress the water as they did the gas.

In this experiment, students will investigate what happens to a gas when the pressure is suddenly released—when it is decompressed.

BACKGROUND INFORMATION

When air is compressed it is heated; when it is released it cools. In the atmosphere, the pressure of the air is reduced when the air rises and is increased when the air falls. This is an important cause of atmospheric temperature changes, and plays a major role in the condensation of water vapor.

175

MATERIAL

- Pressurized can of air freshener (partially used)
- Thermometer
- Stopwatch or wall clock

STUDENT PROCEDURE

When air rises through the atmosphere, the pressure exerted on it by surrounding air molecules is reduced. Students can investigate the effects of reducing pressure by measuring the temperature of gas which is released from an aerosol container. A partially used aerosol can will give more spectacular results than a brand new can. An aerosol can dispenses a combination of liquid contents and gas propellent. When the can is new almost all of the discharge is the liquid content, but as the can is used, more and more gas is expelled with the liquid. It is the temperature of this gas component which we wish to study; hence, used aerosol cans are more appropriate for this experiment than new ones.

1. Spray an aerosol can of air freshener directly on the bulb of a thermometer for one second. *Do not point the can toward another person!* Record the final temperature.
2. Let the thermometer bulb warm back to room temperature. Spray the air freshener on the bulb for two seconds. Again record the final temperature.
3. Let the thermometer warm back to room temperature again. Find the final temperatures when the bulb is sprayed with air freshener for times of three, four, and five seconds. (Always wait until the thermometer has warmed back to room temperature before starting to time!)

ANALYZING RESULTS

Have students graph the temperatures for spray times of zero, one, two, three, four, and five seconds. (The final temperature for a spray of zero seconds is the room temperature.) Most students will find the maximum amount of cooling occurs before five seconds, but the exact nature and amount of this cooling depends upon the nature of the aerosol can. Several deaths have been reported recently from inhaling the spray from aerosol cans. Don't miss the opportunity for a relevant discussion of the results of this experiment.

176

INVESTIGATING FURTHER

You may wish to demonstrate the reverse of this experiment by pumping air into a bicycle tire with a tire pump. In this case, the air pressure is being increased. Have the students feel the barrel of the tire pump.

Have students discuss the role of rising air in the condensation of water vapor. They may want to review the calculation of cloud heights made in Meteorology Experiment 1, "Measuring Dew Point and Humidity."

9

Reading and Plotting
Weather Data

In previous experiments, the student has measured temperature and dew point and has seen a relationship between temperature and decreasing pressure. In this experiment, we shall look at values for these variables over the continental United States to see if we can discover any relationship between the patterns formed. In the process, students will gain practice in reading the standard symbols plotted on weather maps. You may wish to ask students to bring weather reports and maps from local newspapers to class in order to motivate a discussion leading into this experiment. In this experiment and the next, wind speeds are given in miles per hour to conform with U.S. Weather Bureau practice. If you wish to work in the metric system, 1 mi/hr = 1.6 km/hr = .45 m/sec.

BACKGROUND INFORMATION

The following observations may be of help in looking at a weather map and thinking about probable changes which may soon occur.

Usually lows and highs move with the prevailing westerlies at an average speed of about 30 to 35 mi/hr in winter and about 20 to 25 mi/hr in summer. There may be large individual variations from this average, however. A pressure depression usually moves with the same speed and in the same direction as in the past 12 to 24 hours. Strong winds in front of a low will tend to slow its motion. The direction of movement tends to be along lines of equal pressure (isobars), but cutting across lines of equal temperature (isotherms) toward an area of high temperature.

Lows also tend to travel toward the area with the greatest fall in pressure during the preceding three-hour period, while highs tend to move toward the area where the greatest rise in pressure is occurring. A strong high pressure area east of a low will tend to retard the low or deflect it to the right or left. Two lows close together will tend to unite. Cloudiness and precipitation usually accompany a moving depression, but the occurrence of rain also depends upon the topography of the region relative to mountain chains and large bodies of water.

Changes in temperature are largely controlled by the wind, and the wind has a direct relation to the pressure distribution, as will be seen from an analysis of a weather map. Winds are caused by three forces: differences in pressure, the Coriolis force, and a centrifugal force (the latter two caused by the earth's rotation). The resultant direction of motion is along the isobars instead of across them. However, friction and turbulence at the earth's surface deflect the winds so that they are pulled around slightly in the direction of the pressure force (from high to low pressure). As a result, in a region of low pressure the air has an inward curving motion in a counterclockwise direction. From a region of high pressure the air moves spirally outward in a clockwise direction.

MATERIAL

- Large daily weather maps, as shown in Figure M-6 (next page), may be obtained from: Superintendent of Documents, Government Printing Office, Washington, D.C. 20402. The minimum subscription period is three months for $2.40. A single subscription is usually sufficient for a school since it is not necessary that students have identical copies.

STUDENT PROCEDURE

The daily weather map consists of a large Surface Weather Map (the

179

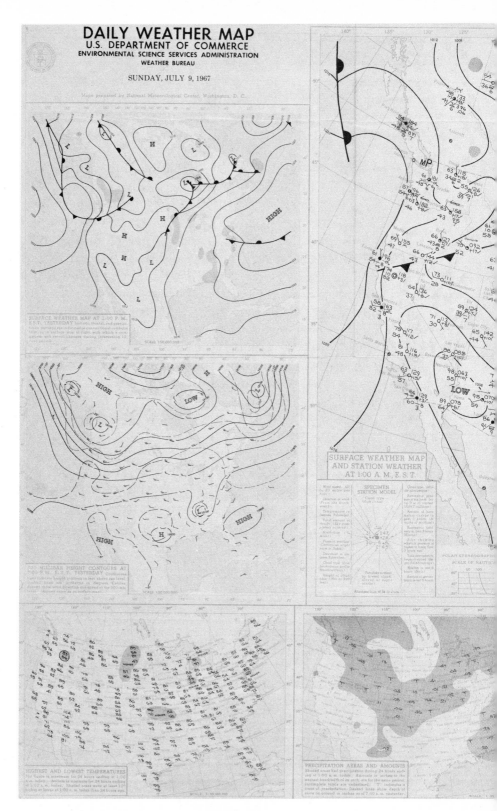

Figure M-6. *The daily weather map consists of a principal surface weather map and five auxiliary maps.*

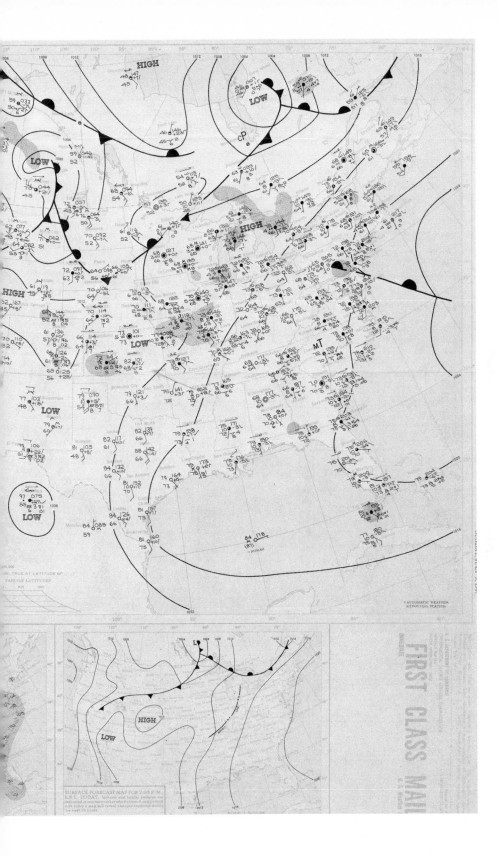

Station Weather at 1:00 AM, EST), and five auxiliary maps. One of these shows the surface weather 12 hours earlier (1:00 PM, EST, yesterday). By comparing these two maps, we can trace actual weather movements.

1. Locate today's surface weather map. Plot each high and low pressure area on the smaller map, which shows yesterday's surface weather. Students can use state borders as guides in locating these centers.

2. Draw arrows from yesterday's pressure centers to today's pressure centers. Circle high pressure areas or low pressure areas which appear to have combined.

Have students fill in Table M-3 from information plotted on the *Surface Weather Map and Station Weather* for what you consider to be an appropriate number of stations (depending upon student ability and available class time).

Table M-3. *Table of weather data to be recorded.*

Station:
Temperature:
Dew Point:
Corrected Barometric Pressure:
Pressure Change in Last 3 Hours and Tendency:
Wind Speed and Direction:
Amount of Precipitation in Last 6 Hours:
Type and Amount of Cloud Cover:
Unusual Additional Notations:

ANALYZING RESULTS

Discuss the wind patterns appearing on different maps. Estimate the farthest distance moved by a pressure center and compute its approximate velocity by dividing this distance by 12 hours. If you have a series of weather maps for several days you may want to have students predict future movements of these centers and then check their predictions against subsequent maps.

INVESTIGATING FURTHER

Have students inspect the map of *Precipitation Areas and Amounts*. Temperatures for most stations can be recorded on this map by reading them from the *Surface Weather Map*. Ask students to then draw lines through stations reporting temperatures of 20°, 30°, 40°, 50°, 60°, etc. Students will have to exercise quite a bit of judgment in drawing such lines. For example, if two adjacent reporting stations reported temperatures of 58° and 62°, one should probably judge that the 60° line passed somewhere between these two stations. In the desert areas of the west, where reporting stations are far apart, the lines may have to be extended horizontally.

Discuss the pattern formed by these isotherm lines. How does it relate to areas of precipitation? Is there a relationship that can be observed between isotherms and isobar patterns?

10

Wind Currents

MOTIVATORS

Your students may have had some difficulty determining wind patterns on the weather map studied during the previous experiment. If so, it may be helpful to keep a daily observation of wind directions and speeds over a period of a week or two. Students may wish to construct an anemometer to measure wind speeds, or may simply estimate a measure of the speed by using the Beaufort scale shown in Table M-4.

The speeds indicated in the table are for a level about 33 ft above ground level. The Beaufort scale is for steady winds, and makes no correction for turbulence or gustiness. If students do not take momentary gusts into account when observing the movement of leaves and trees, then their estimates of wind speed may be too high.

BACKGROUND INFORMATION

Our direct knowledge of the pressure and movements of the upper air comes from observations of cloud movements, from pilot balloons, and from instrument packages sent aloft. At upper levels of the atmosphere

Table M-4. *Beaufort scale of wind speeds.*

Beaufort Number	Wind Description	Speed, mi/hr.	Criteria
0	calm	less than 1	smoke rises vertically
1	light air	1-3	direction shown by smoke drift, but not by wind vanes
2	slight breeze	4-7	wind felt on face; leaves rustle; vane moved by wind
3	gentle breeze	8-12	leaves and small twigs in constant motion; wind extends light flag
4	moderate breeze	13-18	raises dust and loose paper; small branches are moved
5	fresh breeze	19-24	small trees in leaf begin to sway
6	strong breeze	25-31	large branches in motion; whistling heard in utility wires
7	moderate gale	32-38	whole trees in motion; inconvenience felt in walking against wind
8	fresh gale	39-46	breaks twigs off trees; generally impedes progress
9	strong gale	47-54	slight structural damage occurs; roof shingles blown free
10	whole gale	55-63	trees uprooted; considerable structural damage occurs
11	storm	64-75	widespread damage; very rarely experienced
12	hurricane	above 75	

the pressure pattern is much simpler than in the surface layers. The isobars of the pressure pattern tend to generally run in an east-west direction. With practically no friction in these upper levels, the principal air movement is from west to east.

MATERIAL

- Helium-filled toy balloons

- Weights (large metal washers, or scrap metal)

- Aluminum foil

- Postcards

- Magnetic compass

- Stopwatch, or watch with a sweep second hand

- City and county maps (optional)

STUDENT PROCEDURE

What is the speed of the wind above the ground? Students can find out by releasing a helium-filled balloon and estimating its velocity. Follow this procedure:

1. Prepare the balloons in the classroom by taping a piece of aluminum foil and a postcard to each of them. The aluminum foil is to aid visibility. The postcard should be addressed to your class, and request that the finder inform you of the time and place the balloon was recovered.

2. Tie weights to the balloon so that it rises slowly when released (about 15 cm/sec). The purpose of this experiment is to measure the horizontal velocity of the balloon—if it rises too rapidly it will soon be so high that its horizontal position cannot be estimated.

3. Release the balloon and start the stopwatch. The observers at the far side of the schoolyard should yell "stop" when the balloon is overhead, and the elapsed time should then be read from the stopwatch and recorded.

4. Measure the distance from the point of release to the point of overhead observation. Very observant students may decide that the distance moved by the balloon and the distance measured along the ground are really not the same. Since the balloon is rising as well as moving horizontally, this is true. But the horizontal part of this motion is the part which is due to the wind, and that is the part which would be measured along the ground.

5. Sketch a rough map of the schoolyard and record the direction of the balloon's flight. A magnetic compass may be used to increase the accuracy.

6. Repeat the procedure with additional balloons to obtain average times and directions. If the balloon rises too rapidly or too slowly, change the number of weights added to the next one.

ANALYZING RESULTS

Have students calculate the average horizontal velocity of each balloon by dividing the measured distance by the time of flight measured on the stopwatch. Obtain average values for the wind velocity.

INVESTIGATING FURTHER

Using the average velocity calculated for your set of balloons, find the distance a balloon would travel in the next 6-, 12-, and 24-hour period if it were to continue to move with its original speed and in its original direction. Discuss how these points could be plotted on the city or county map by converting from actual distances to scale distances. Use the scales shown on the maps to actually plot the estimated points.

If any of the postcards which were attached to the balloons are returned to the class, locate the point of interception on a map. Compare this location to the predicted line-of-flight. Discuss factors which might help account for differences, including subsequent weather reports.

11

Vertical Patterns: Pressure, Temperature, and Dew Point

MOTIVATORS

In the previous experiment, balloons were used to investigate wind currents above the ground. If other measuring instruments are attached to the balloon, such variables as pressure, temperature, and dew point can also be measured. The measurement of pressure varies so regularly with height that it can be taken as an indirect approximate measure of the distance of the balloon above the ground. Table M-5 relates the height in kilometers to the pressure in millibars and the temperature in degrees Celsius under standard atmospheric conditions.

The most striking thing about the table is, of course, that the temperature does not continue to drop with increasing elevation. This surprising fact was discovered with the use of sounding balloons, and confirmed as a generalization between 1899 and 1902. The nearly isothermal region (which extends from about 11 to 95 km) is called the *stratosphere*.

BACKGROUND INFORMATION

If a parcel of air is raised through the atmosphere, its temperature will drop because of the decreasing pressure upon it (as in Meteorology Experiment 8). This temperature change occurs without addition or

Table M-5. *Variation of atmospheric pressure with height.*

Height, m	0	1000	2000	3000	4000	5000	6000	7000	8000
Pressure, mb	1013	903	795	700	616	540	472	415	356
Temp., °C	15	8.5	2	−4.5	−11	−17.5	−24	−30.5	−37

Height, m	9000	10,000	12,000	14,000	16,000	18,000	20,000
Pressure, mb	308	264	235	193	141	103	55
Temp., °C	−43.5	−50	−55	−55	−55	−55	−55

removal of any heat—such a temperature change is termed *adiabatic*. (If we imagine a parcel of air being lowered through the atmosphere to greater pressures, a temperature rise will occur). From the physical properties of gases, it can be mathematically shown that when dry air rises above the ground surface, the cooling due to expansion under lower pressures is 1°C per 100 m. If there is considerable moisture present in the rising air, some of this moisture may begin to condense as the air cools, releasing latent heat of condensation. The net result is that the cooling rate appears to be much slower for "wet" air. The rate of cooling for saturated air is not as constant as the rate for dry air, depending not only upon the change in elevation, but also upon temperature, the original pressure, and whether most of the condensed moisture is carried along with the rising air or precipitated out. Thus while the dry adiabatic rate is always 1°C/100 m, the saturated or wet adiabatic rate can vary from 0.4°C to nearly 1°C/100 m.

In actual practice, however, overlying air does not always grow colder at these adiabatic rates. Air is constantly gaining and losing heat by radiation, absorption, and conduction, as well as by evaporation and condensation. In addition, horizontal air movements may bring an influx of warm or cold air from other sources. Thus the real vertical distribution of temperature is often quite different than that caused by adiabatic processes. The actual change of temperature with elevation is called the *lapse rate* of the air. Ordinarily, the lapse rate of air is less than the dry adiabatic rate and about equal to the wet adiabatic rate. Normally the lapse rate is less when the pressure is high than when it is low, and less in winter than in summer.

Plotting temperatures as they vary with height (as measured by pressure) can tell us about the stability of the air when we compare its lapse rate with the adiabatic rate. Let us first suppose that the wet or dry adiabatic rate is greater than the actual lapse rate. If we lifted some of this air to a higher elevation, cooling it at the adiabatic rate it would be colder than the air surrounding it and therefore more dense, tending to sink back to the ground. In this situation the air is said to be stable, for if displaced it tends to return to its original position. When the lapse rate of air is less than the wet or dry adiabatic rate, the air is stable.

It is possible that the air may actually grow warmer with higher elevations. Such a condition is called an *inversion*. If we lift some of this air to a higher elevation, cooling it adiabatically, it would be much, much colder than the surrounding air and would sink. An inversion of air means extreme stability. The lower air cannot rise, but is confined to its own level as if it were covered by a "ceiling" or "lid." When this air is calm, smoke and gas fumes become trapped and form smog. Large-scale downward motion in the atmosphere is called *subsidence*. If subsidence occurs when the lapse rate is smaller at higher elevations than at the ground, the top part of the subsiding air can be warmed by adiabatic compression more than the bottom part, resulting in a temperature inversion. Temperature inversions are also produced when the ground is colder than the overlying air, since this condition will cool the lower air faster than the upper air.

Suppose that the dry adiabatic rate is less than the actual lapse rate of the air. If we lift some of this air to a higher elevation we would find that the surrounding air was colder than the adiabatically cooled air. The "foreign" parcel of air, being warmer and less dense, would continue to rise. Likewise, air starting downward would not warm as fast as the surrounding air and hence would continue to sink. In this case the air is unstable, for given a push in either direction it will continue to go in that direction. If there is a cold current in the upper air, rising unstable air will continue through it and be cooled below its dew point, resulting in cloudiness and rain. Thus stable air favors fair weather, and an unstable condition is conducive to cloudiness and rain.

Of course, as rising air cools closer and closer to its dew point, it becomes more and more saturated in terms of its potential to hold water vapor. Since the wet adiabatic rate is less than the dry adiabatic rate, such air can change from stable to unstable if some other process causes it to rise. Such air is conditionally unstable and a potential source of danger.

But vertical patterns in the air can tell us more than just its stable or unstable nature. Measurement of dew point tells us about the moisture content of the air. With this additional information it may be possible to

190

identify the source of the air mass as cold or warm (polar or tropical) and wet or dry (maritime or continental). The usual characteristics of the different air masses are:

- *Polar Continental (cP):* Originates over an ice- or snow-covered surface, which aids in cooling at the bottom. Therefore, the temperature often increases from the ground up to a considerable elevation. This temperature inversion results in marked stability.

- *Polar Maritime (mP):* Originates usually as cP air in Siberia, but moving eastward over relatively warm Pacific waters it becomes warm and humid in its lower levels. Thus, it develops a steep lapse rate and conditional and convective instability.

- *Tropical Maritime (mT):* Originates in the Pacific between Baja California and Hawaii or in the Gulf and Caribbean regions. Subsidence is characteristic, but the surface layer may be cool and moist so that conditional instability may be present, especially in summer.

- *Tropical Continental (cT):* Originates over northern interior Mexico and the arid southwestern United States. Because there is intense heating at the surface, this air is turbulent with convection as high as 2 mi. The lapse rate approximates the dry adiabatic rate, but the air remains cloudless because of its extreme dryness. No cT air reaches North America in the winter.

MATERIAL

- Graph paper

STUDENT PROCEDURE

For each of the four sets of typical air-mass soundings shown (Tables M-6 to M-9), have students plot temperature vs. pressure and dew point vs. pressure on a single graph. Plot decreasing pressure vertically (to indicate an increase in height) as shown in Figure M-7, next page. Lightly shade the area on the graph between the temperature and dew point curves for each air mass.

Table M-6. *cP air at Fairbanks, Alaska, in winter.*

Pressure, mb	995	960	928	908	868	855	828	753	745	615	525
Temperature, °C	−26	−19	−13	−13	−12	−8	−8	−13	−13	−22	−31
Dew Point, °C	−27	−21	−17	−18	−19	−19	−19	−26	−26	−33	−43

Table M-7. *mT at Miami, Florida, in winter.*

Pressure, mb	1020	1010	880	750	625	508	460	400
Temperature, °C	19	21	12	8	0	—10	—15	—22
Dew Point, °C	15	17	10	—6	—14	—25	—29	—34

Table M-8. *mP at Washington, D.C., in winter.*

Pressure, mb	1010	928	870	850	715	578	470	400
Temperature, °C	4	1	2	2	—6	—18	—27	—35
Dew Point, °C	2	0	1	—3	—7	—21	—28	—46

Table M-9. *cT at El Paso, Texas, in summer.*

Pressure, mb		880	860	770	700	615	540	400
Temperature, °C		34	34	26	20	9	—2	—17
Dew Point, °C		—4	—5	—9	—12	—21	—27	—46

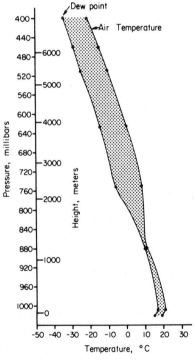

Figure M-7. *A student's graph of the data in Table M-6.*

ANALYZING RESULTS

Ask students to identify temperature inversions, sections where the lapse rate is greater than the dry adiabatic rate, and sections where the lapse rate is less than the dry adiabatic rate.

Figure M-7 shows the sounding-data graph for the air mass data given in Table M-6. Note the temperature inversion between 1020 mb and 1010 mb. Heights have been marked on the pressure scale using the information in Table M-5. Between 0 and 1000 m, the temperature falls from about 21°C to 13°C, giving an average lapse rate of approximately 0.8°C/100 m. This is less than the dry adiabatic rate of 1°C/100 m, and indicates that the air is stable between these levels. The humidity of the air is higher where the temperature and dew point lines are closer together. Because the dew point temperature is approximately equal to the air temperature at a height of about 1200 m, we might expect to find clouds at that altitude. Above this level, the air becomes progressively dryer.

INVESTIGATING FURTHER

Have students plot data from the following two soundings. Using the graphs made previously, discuss the problems of identifying these two unknown air masses as cP, mP, cT, or mT.

Table M-10. *Unknown air mass, number 1.*

Pressure, mb	962	900	850	768	741	700	672	622	584	536
Temperature, °C	—4	—7	—6	—19	—9	—12	—14	—20	—20	—24
Dew Point, °C	—5	—9	—8	—12	—12	—16	—19	—23	—28	—38

Table M-11. *Unknown air mass, number 2.*

Pressure, mb	996	898	850	840	794	724	700	692	642	626
Temperature, °C	14	11	8	8	7	2	—1	—2	—6	—6
Dew Point, °C	13	11	6	6	—13	—11	—9	—9	—14	—23

12

Patterns on
Constant-Pressure Maps

MOTIVATORS

The adiabatic charts drawn in the previous experiment give information about an air mass along a single vertical line. How can data from soundings along different vertical lines be related and mapped? The most common way to do this is with a constant-pressure map (also known as a pressure-contour map). In this experiment, students will construct a constant-pressure map from actual U.S. Weather Bureau data. We have left the data in its original units (pressure in millibars and height in feet) which will make it compatible with any similar data you might wish to obtain from the Bureau. If you wish to work in the metric system, simply multiply all heights by 0.3 (1 ft = 0.3 m).

BACKGROUND INFORMATION

A constant-pressure map is drawn by choosing a fixed pressure level such as 850, 700, 500, or 300 mb. (Often a separate map is constructed for each of these levels.) The height, in feet, at which this pressure occurred is then plotted at the position of each reporting station. Contour lines are drawn connecting points of equal elevation. These contour lines show the horizontal variation of pressure. Low contour lines indicate areas of low pressure in relation to height (since if we were to rise to the higher surrounding heights the pressure would be less as we went up).

At altitudes of 18,000 to 20,000 ft, the highs and lows and the irregularities of isobars which are so common at the surface give way to a smooth, wavelike succession of troughs of low pressure and ridges of high pressure. There is a mathematical relation between these wave lengths and the speed of advance, so that the movements of these troughs and ridges can be predicted with fair accuracy for several days in advance. In general, weather disturbances at the earth's surface move along the isobars between the principal troughs and ridges at a rate proportional to the pressure differences between them. Thus constant-pressure charts are very useful for preparing two- to five-day advance weather forecasts.

Isotherms and dew point lines may also be recorded on constant-pressure maps.

MATERIAL

- Maps showing the locations of U.S. Weather Bureau Stations. (The small *Precipitation Areas* or *Highest and Lowest Temperatures* maps which are part of the daily map may be used. See the materials section for Meteorology Experiment 9.)

- Red and green pencils

STUDENT PROCEDURE

Table M-12 (next page) gives the heights at which the atmospheric pressure was 850 mb on January 21st. Students can make a constant-pressure map by following these steps:

1. Write the heights where the pressure was 850 mb beside the reporting stations on a map. Draw a contour line through points where the height was 14,300 ft. Draw additional lines for heights of 14,500, 14,700, 14,900, and 15,100 ft. Label the circular areas of high and low pressure.
2. Write the temperatures beside the proper reporting stations. With red pencil, draw lines of equal temperature (isotherms) connecting temperatures of 4°, 2°, 0°, −2° and −4°. Label the warm and cold air centers W and C.
3. Write the dew point temperatures beside appropriate reporting stations. With a green pencil, draw lines of equal dew points (isodrosotherms) for dew points of 10°, 5°, 0°, −5° and −10°. Mark a green M in centers of high dew point, and a green D in centers of low dew point.

Table M-12. *850-mb data for January 21st.*

STATION	HEIGHT	TEMP.	DEW POINT
Portland	14,340	—3.2	—9.5
Albany	14,350	—2.2	—18.5
Nantucket	14,600	—0.2	—16.5
New York	14,560	2.5	—15.5
Washington, D. C.	14,630	2.2	0.5
Norfolk	14,940	3.2	0.2
Hatteras	15,030	3.5	—6.2
Charleston	14,760	8.5	7.5
Jacksonville	14,710	9.8	7.5
Tampa	14,820	11.8	11.8
Miami	15,050	15.2	12.9
Mobile	14,190	10.4	6.0
Montgomery	13,900	8.2	6.5
Atlanta	14,330	7.5	5.8
Raleigh	14,590	5.5	5.2
Pittsburgh	14,290	0.8	—9.2
Buffalo	14,190	—1.2	—14.8
Detroit	13,940	—3.5	—7.5
Columbus	13,820	1.5	1.2
Nashville	13,260	7.2	5.5
Jackson	13,630	—2.2	—3.8
New Orleans	14,250	6.8	—13.5
Lake Charles	13,910	3.2	—14.2
Shreveport	13,490	—4.8	—6.2
Little Rock	12,950	—4.8	—6.5
St. Louis	12,750	—2.5	—4.2
Springfield	13,240	—1.2	—2.8
Nassau	13,650	—11.8	—14.2
Omaha	13,560	—9.8	—14.5
Kansas City	13,130	—7.2	—10.5
Oklahoma City	13,620	—6.5	—8.2
Fort Worth	13,900	—6.2	—9.2
San Antonio	14,420	0.8	—18.2
Corpus Christi	14,560	2.2	—7.8
Brownsville	14,810	4.8	1.2
Midland	14,510	—2.2	—9.2
Amarillo	14,130	—3.5	—12.5
Dodge City	13,900	—6.8	—8.2
Pueblo	14,320	——	——
North Platte	13,870	—7.2	—10.2
Rapid City	13,930	0.2	—12.5
Bismarck	13,820	—2.5	——

ANALYZING RESULTS

Ask students to discuss patterns of pressure (height), temperature, and moisture (dew point).

INVESTIGATING FURTHER

On a separate map, plot the data shown on Table M-13 for 500-mb pressures. Compare the pressure, temperature, and dew point patterns on the 850-mb and 500-mb maps.

Table M-13. *500-mb data for January 21st.*

STATION	HEIGHT	TEMP.	DEW POINT
Portland	15,520	—18.5	—31.5
Albany	15,550	—19.5	—21.5
Nantucket	15,560	—16.8	—25.2
New York	15,610	—17.2	—19.5
Washington, D. C.	15,620	—15.2	—17.8
Norfolk	15,660	—14.2	—15.5
Hatteras	15,690	—14.2	—17.2
Charleston	15,680	—12.2	—14.8
Jacksonville	15,720	—11.1	——
Tampa	15,730	—11.8	—13.2
Miami	15,810	—11.5	—17.2
Mobile	15,660	—12.0	——
Montgomery	15,620	—12.3	——
Atlanta	15,660	—13.2	—17.8
Raleigh	15,640	—13.2	—16.2
Pittsburgh	15,570	—15.8	—17.2
Buffalo	15,530	—19.8	—22.2
Detroit	15,510	—19.8	—28.2
Columbus	15,530	—17.2	—22.2
Nashville	15,510	—15.5	——
Jackson	15,510	—14.5	——
New Orleans	15,650	—14.7	——
Lake Charles	15,510	—14.5	——
Shreveport	15,400	—26.8	——
Little Rock	15,330	—28.6	—36.2
St. Louis	15,390	—19.8	—25.5
Springfield	15,460	—16.8	—21.5

Nassau	15,450	—21.2	——
Omaha	15,420	—21.8	——
Kansas City	15,360	—21.5	—28.2
Oklahoma City	15,350	—28.2	——
Fort Worth	15,420	—26.8	——
San Antonio	15,560	—18.3	——
Corpus Christi	15,610	—17.6	——
Brownsville	15,690	—15.7	——
Midland	15,540	—20.9	——
Amarillo	15,420	—27.0	——
Dodge City	15,380	—28.2	—37.8
Pueblo	15,500	—24.2	—33.5
North Platte	15,430	—24.5	—36.5
Rapid City	15,470	—23.8	——
Bismarck	15,480	—22.0	——

Section Four

ASTRONOMY

1

Changes in the Sky

MOTIVATORS

The unifying theme of the astronomy experiments is the idea of change. What changes can a student observe when he looks into the sky? Unfortunately, most of the celestial changes are too slow to observe with the unaided eye. You may wish to have the class make a collective list of sky changes they have observed over longer periods of time. This list could include such changes as the daily motions of the sun and moon, phases of the moon, the changing length of daylight, the change of the position of the sun in the sky at a given time of day, and the motion of star constellations.

BACKGROUND INFORMATION

In this experiment, students will measure apparent changes in the relative positions of stars in the sky. These apparent motions are recorded by means of a photograph exposed to the sky for 90 minutes (see Figure A-1, next page). The intent of the experiment is to introduce the idea of angular displacement as well as the more familiar linear displacement, and to show that the values of these displacements are relative to the point of observation.

201

Figure A-1. *Changes in the sky during a time exposure of 90 minutes. Fritz Goro,* LIFE Magazine © *Time, Inc.*

MATERIAL

- String
- Ruler
- Protractor
- Copies of the photograph tracing (Figure A-2)

STUDENT PROCEDURE

Figure A-1 is a 90-minute time exposure of the sky on a summer night in the northern hemisphere. A camera with its shutter open was fixed on the radar antenna. The most obvious thing about the photograph is the apparent motion of the stars with respect to the antenna during the 90-minute period.

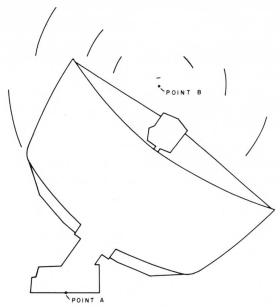

Figure A-2. *A tracing of eight of the star trails in Figure A-1.*

Figure A-2 is a student's tracing of eight of these star trails in relation to the antenna. With such tracings students can measure how far the stars move during the 90-minute exposure. There are at least three ways in which this can be done:

1. Measure the length of each trail using a ruler. If the trail is curved you may wish to measure with a piece of string, which can then be straightened out along the ruler.
2. Imagine that you are standing at the bottom of the photograph. In Figure A-2 we have taken the center of the base of the antenna as such a point (Point A). Draw lines from this point to both ends of of each star trail. Find the apparent angle that each of the stars moved through during the 90-minute exposure.
3. There is a "natural" center which appears in all such photographs. This center is marked "Point B" in Figure A-2. Imagine yourself standing at Point B. Draw lines from this point to both ends of each star trail. What apparent angle did each of the stars move through during the 90-minute exposure?

ANALYZING RESULTS

Have students record the three measurements of the displacement of each of the eight stars considered. Ask students which star appeared to move farthest during the 90 minutes, and which star appeared to move the least. The answer to this question will depend upon which of the three methods are compared.

Students may feel that one of these methods is "right" and that the other two are somehow wrong. Method one is limited because the photograph is obviously scaled down from the actual distances involved, and we have no information about the "shrinkage factor." Figure A-3 is a photograph of 1-ft rulers placed at various distances from the camera. Have students measure the actual lengths of the images of these rulers on the photograph. They should discover that rulers in the background are shrunk more than objects in the foreground. Therefore, the problem is more complicated than finding a single "shrinkage factor"; one would have to know the "shrinkage factor" for each of the different distances of the objects in front of the camera. In terms of the star trails, this means that we would have to know how far away the stars are.

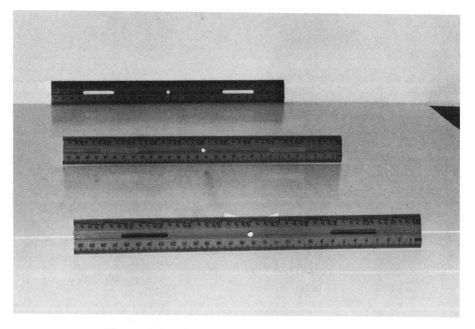

Figure A-3. *The rulers in this photograph are 3, 4, and 5 ft away from the camera lens.*

Since Figure A-3 tells us how far each ruler was from the camera, we can look for a relationship between the "shrinkage factor" and the distance from camera to ruler. Give a xerox copy of the photograph to each student. We can compute the "shrinkage factor" by the simple formula:

$$\text{shrinkage factor} = \frac{\text{measured length of image}}{\text{actual length of object}} \quad .$$

Make a graph of shrinkage factors vs. distance to the camera. Use the graph to estimate what the shrinkage factors might be for other distances.

Have students discuss how the angles in method two compare with those for the same star trail measured by method three. Ask what would happen if they chose another "Point A."

INVESTIGATING FURTHER

If you have some students that are interested in photography, your class might want to consider whether the use of different cameras would affect the shrinkage factor vs. distance graph. This investigation could be developed into a nice science fair project.

The differences in angle values that arise between methods two and three should give rise to the idea of relative motions. Ask the students if they can be sure that the stars even moved at all. Have them imagine that they are riding a merry-go-round with a transparent top. What kind of picture could be taken from such a position?

2

The Angular Velocity
of the Sun

MOTIVATORS

In the previous experiment, students noticed the circular nature of star motions and measured this motion in terms of angular displacement. In this experiment, they will measure the angular velocity of the sun. Challenge students by asking them to guess whether the angular velocity of the sun is greater or smaller than the average value found for the stars.

BACKGROUND INFORMATION

The basic idea behind measuring the angular velocity of the sun is to take two sightings at definite time intervals apart and determine the angular difference in the direction of these two sightings. *Don't sight the sun directly,* however, since it could damage your eyes. We can "sight" it indirectly by looking at the shadows it casts. As the sun moves across the sky, the shadows it casts move across the ground. We can measure the sun's movement by measuring the movement of these shadows. Figure A-4 shows the relationship of these motions. Notice the two imaginary lines drawn from the sun to the top of the post and extended to the ends of the shadows. It is the angle formed by the intersection of these two imaginary lines which we need to measure to get the angular displacement of the sun.

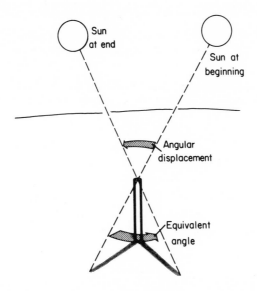

Figure A-4. *The angular displacement of the sun can be measured by tying string from the top of a post to the end of the post's shadow at two different times.*

MATERIAL

- Post or table (about 1 m high)
- 2 stakes
- String
- Large protractor

STUDENT PROCEDURE

Students can measure the angular velocity of the sun by measuring the angular velocity of a shadow which the sun casts.

1. Locate a fence post in the schoolyard whose top you can reach. If this is not convenient, you can use one of the legs of a table, provided the table is not moved during the experiment.

2. Have students find the end of the shadow cast by the post on the ground. Drive a small stake in the ground at that point.

3. Wait for exactly one-half hour. Notice the new position of the shadow end, and mark it by driving the second stake in the ground.

4. Locate the lines from the top of the post to each of the stakes you have driven by tying string between these points.

5. Use a large protractor to measure the angle between the strings (a blackboard protractor works well). This is the angular displacement of the sun in one-half hour. What would be the angular displacement of the sun in one hour? What is the angular velocity of the sun in degrees per hour?

ANALYZING RESULTS

Compare the values for the apparent angular velocity of the sun which different student groups obtain. How do these values compare with the angular velocities obtained in Astronomy Experiment 1?

INVESTIGATING FURTHER

Have students drive stakes in the ground to plot the position of

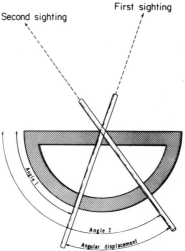

Figure A-5. *A measuring instrument for direct sightings can be made from a plastic straw and a protractor. The protractor must not be moved between sightings.*

the shadow every half hour for four or five hours. What kind of line do the stakes form? Ask students whether the value of angular velocity would vary depending upon the time of day the measurements were taken.

On a night when there is a full moon, it is possible to see shadows cast by the moonlight (in an area not illuminated by street lights). In such a case, it is possible to measure the angular velocity of the moon using the same method. Ask students to volunteer to gather such measurements for the class.

Unlike the sun, the moon can be sighted directly. A simple sighting device can be made from a plastic straw and a protractor, as shown in Figure A-5. Other students may wish to obtain measurements by using this device as Figure A-5 suggests. The protractor should be taped to a post so that it is not moved between sightings. Have students compare the angular velocities of the sun and moon.

3

Angular and Linear
Velocities

MOTIVATORS

In previous experiments, the velocities of objects in the sky have been measured as angular velocities. Are these velocities related to the ordinary linear velocities measured in miles per hour here on earth? In this experiment, students will measure both the angular and linear velocities of a moving object and compare the results.

BACKGROUND INFORMATION

This experiment is designed to show that the relationship between angular and linear velocity depends upon the distance from the observer to the moving object. This can be demonstrated experimentally without appealing to involved mathematical arguments. Older students with sufficient mathematical background may see the mathematical connection if they know about the radian measure of angles. Any angle can be placed with its vertex on the center of a circle. The measure of the angle θ, in radians, is defined to be the ratio of the arc length, s, intercepted by the sides of the angle on the circle to the radius of the circle, r. That is,

$$(\theta) = \frac{s}{r} \ .$$

If, however, the measure of the angle in radians is already known, the length of the arc, s, can be found by the formula

$$s = r(\theta).$$

It is this arc length which we must know to compute the linear velocity of a distant object.

The conversion of angle measure from degrees to radians is relatively simple. An angle of 360° would intercept the entire circumference of the circle, $2\pi r$. So, $360° = 2\pi r/r$ radians, or $1° = 0.018$ radians.

Thus to find the linear velocity, we must know the angular displacement in radians, and r—the distance from the observer to the object.

This experiment avoids the preceding mathematics by analyzing the data with a scale drawing, a technique which will be used again in subsequent experiments. The teacher may wish to include the above discussion in classes where the students have had sufficient mathematics.

MATERIAL

- Moving coffee can
- Meter stick
- Protractor
- Stopwatch or wall clock

To make a moving coffee can, you will need an empty 1 lb coffee can, two plastic lids, two toothpicks, a rubber band, wire, and a weight. Remove both metal ends of the can with a can opener. Punch a hole in the center of each of the plastic lids. Attach the weight tightly to one strand of the rubber band with the wire. Figure A-6 shows how to assemble the parts. Each end of the rubber band is pushed through a plastic lid and looped around a toothpick.

Figure A-6. *The assembly scheme for constructing a moving coffee can.*

211

To use the coffee can, roll it on the floor so that the weight will twist and tighten the rubber band. If the band is wound tightly enough, the can will move back to its original position under its own power when it is released.

STUDENT PROCEDURE

Students can find a relationship between angular and linear velocities by measuring the velocity of a moving object in the laboratory by both methods. For the moving object, use a moving coffee can.

1. Tape a sheet of paper on the floor, at least 2 m away from the coffee can and to one side of its path of travel. Mark a point near the center of this paper. This point will be the observer's position.
2. Place a meter stick on edge so that one end of the meter stick is on the point marked on the paper. Sight down the meter stick; move it until it is in line with the coffee can. Mark this observation line on the paper.
3. Measure the distance from the observer's position (point) to the coffee can with the meter stick. Record this distance.
4. Have your partner let go of the coffee can and let it roll for five seconds. Catch the can and hold it in its final position. (If the can moves rapidly, you may need to shorten the time accordingly.)
5. Sight along the meter stick to the final position of the can. (Be sure one end of the meter stick still rests on the point marked on the paper as the observer's position.) Mark the new line of sight on the paper. Remove the meter stick. The angle between the two lines of sight is the angular displacement of the coffee can in five seconds.
6. Measure the distance from the mark to the final position of the coffee can. Record this distance.
7. Measure the linear distance the coffee can moved with the meter stick. It is this distance which we want to relate to the angular displacement marked on the paper.

Before analyzing your data, repeat the experiment, but this time move back so that the observer's position marked on the new sheet of paper is about 4 m away from the coffee can. Again find the line of sight and the distance to the initial position of the coffee can. Let the can move, and then find the line of sight and the distance to the final position of the can, and the distance actually moved by the can.

ANALYZING RESULTS

After repeating the experiment, students will begin to realize that the

212

size of the angular displacement depends upon the distance from the observer to the object.

Can the angular displacement be used to calculate how far the object moved? If the distances from the observer to the object at the beginning and end of its motion are known, the answer is yes. The easiest way to do it is to make a scale drawing of the situation. Keep all angles the same, but divide all distances by 20. Each student should be able to make a scale drawing similar to Figure A-7. The distance between the initial and final positions of the can may be measured from such a drawing. Since this is a scale drawing, the measurement must be multiplied by 20 to obtain the value for the actual distance.

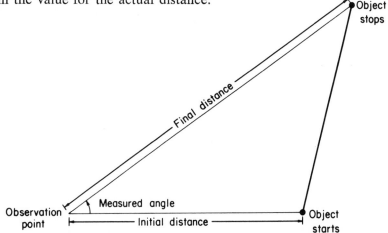

Figure A-7. *The distance an object moves can be measured indirectly by constructing a scale drawing similar to this one.*

How does the calculated distance compare with the actual measurement?

INVESTIGATING FURTHER

Ambitious students may wish to derive an algebra formula relating linear displacement to angular displacement and the distance between the object and the observer. If the coffee can is always moved so that its initial and final positions are equidistant to the observation point, the relation can be graphed in the following way. Measure the ratio of linear displacement to angular displacement over several trials in which the distance from the observer to the object is varied. Graph this ratio against the observer-object distance.

In Astronomy Experiment 2, students measured the apparent angular velocity of the sun. If the distance from the earth to the sun is known to be 150 million km, students should be able to construct a scale drawing to estimate the apparent linear velocity of the sun. A convenient scale to use is 1 cm = 8 million km. They should obtain an answer of approximately 39 million km/hr (24 million mi/hr). At this point, however, students have no way of knowing whether this is actual or only apparent velocity.

4

The Distance to the
Moon by Scale Drawing

MOTIVATORS

In the previous experiment students found a way to calculate the linear velocity of an object in the sky, provided the angular velocity of the object and the distance between the object and the point of observation (earth) were known. In Astronomy Experiments 1 and 2 students measured angular velocities. It should be apparent that what is needed now is a method for measuring distances from the earth to nearby celestial objects like the sun and moon. In this experiment, we shall present data which will allow the student to calculate the distance from the earth to the moon by construction of a scale drawing.

BACKGROUND INFORMATION

The earliest measurements of the distance from the earth to the moon were made by a method known as triangulation. If an object is sighted from both ends of a base line so that the angles between the lines of sight and the base line are known, together with the length of the base line, the distance to the object can be calculated. This calculation is relatively straightforward if the student knows trigonometry. However, the distance can easily be found without the use of trigonometry by constructing a scale drawing much like the method used in the previous experiment.

The success of triangulation depends upon the size of the base line in relation to the distance to be measured. The more distant the object being sighted, the longer the base line must be. Otherwise the two sight lines are so nearly parallel that differences in the angles are less than the errors of measurement.

An early measurement of the distance to the moon was made by triangulation in 1752. One end of the base line was an observatory in Berlin; the other end was an observatory in South Africa. This is a surface distance of about 10,000 km, or a straight-line distance of 9040 km.

MATERIAL

- Ruler

- Protractor

STUDENT PROCEDURE

An early measurement of the distance to the moon was discovered in 1752 by making two observations. Using the measurements made then, students can construct a scale drawing which will give a value for the distance from the earth to the moon.

The moon was sighted simultaneously from an observatory in Berlin and from an observatory in South Africa. In Berlin, the moon appeared to be 21° away from the zenith (a vertical line pointing directly overhead). In South Africa (at the same time) the moon appeared to be much closer to the horizon—the moon was 70° from the zenith (overhead).

To make a scale drawing we need to know the distance between the two observatories. They were approximately 10,000 km apart. However, this distance is measured on the surface of the earth; it is obviously not a straight-line distance. The radius of the earth is about 6400 km, and the earth's circumference is 40,000 km. Thus, the two observatories were separated by about ¼ of the earth's circumference.

Students should be able to complete the scale drawing using the following procedure:

1. Use a scale of 1 cm = 1000 km. Draw a circle to scale representing a cross section of the earth. (The radius of the earth is approximately 6400 km).

2. Mark a point on the circumference of the circle to represent the Berlin observatory. Draw the zenith line at Berlin. (The imaginary line point- ing overhead would pass through the center of the earth if extended in the other direction.)

3. The observatory in South Africa was about ¼ of the earth's circumfer- ence away. What angle would its zenith line make with the Berlin zenith line where they meet at the earth's center? (There are 360° in a circle.) Measure the proper angle, and draw the zenith line for the South African observatory.

4. Draw the line of sight to the moon at Berlin by constructing an angle of 21° with the Berlin zenith at the earth's surface.

5. Draw the line of sight to the moon at South Africa by constructing an angle of 70° with the South African zenith at the earth's surface.

6. Extend the two lines of sight until they meet. (Tape additional sheets of paper onto your original drawing until you can locate a point of intersection.)

7. Measure the length of the sight lines, remembering that 1 cm = 1000 km. How far away is the moon?

ANALYZING RESULTS

Have students compare their answers. Figure A-8 shows how student drawings should appear. Some will want to repeat their construction in an attempt to improve their accuracy. The entire class should be able to arrive at a "best value" after many results have been compared.

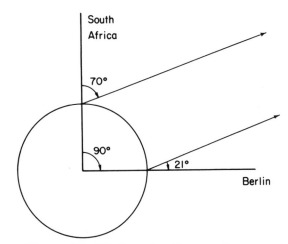

Figure A-8. *Finding the distance from earth to moon begins with this scale drawing.*

The average distance of the moon from the earth is about 384,000 km. A relative error of only 1% will give a variance of 3840 km in the final answer. On the scale drawing, this means an error of only 3.8 cm. Help students realize that an error of 3840 km in a total distance of 384,000 km is not nearly as bad as an error of 3840 km in, say, 10,000 km (the distance between the observatories).

INVESTIGATING FURTHER

You may want to have students investigate ways of calculating the sides of the triangle by trigonometric means. The straight-line distance between the observatories would be approximately 9040 km.

Ask students if the moon was closer to Berlin or to South Africa according to their drawing.

In Experiment 2, students found the angular velocity of the sun and moon. The angular velocity of the moon is about $13°/hr$. We can now find its apparent linear velocity (relative to the earth) with another scale drawing. Draw an isosceles triangle with sides of 384,000 km (to an appropriate scale) which are separated by an angle of $13°$. Measure the base line, and convert to actual distance to find the apparent distance the moon moves in one hour. Your answer will be more than 80,000 km/hr (50,000 mi/hr). Does the moon really move this rapidly, or is this figure an illusion? We shall find out in subsequent experiments.

5

The Distance to the
Moon by Radar

MOTIVATORS

In the previous experiment, students calculated the distance from the earth to the moon from data taken from astronomical sightings. Today radar is used to measure this distance. This is an analogy experiment which explains the reasoning behind the radar measurement and subsequent calculation.

BACKGROUND INFORMATION

Shortly after World War II, scientists realized that radar which had been developed to track and measure the distances to airplanes might also be used to measure the distance to the moon and nearby planets. Radar is a highly focused beam of ultra-high-frequency radio waves which travel with a constant velocity and in a straight-line path. When these waves reach a stationary solid object, they are reflected back at the same velocity they were emitted. Thus, the time between their emission and reception multiplied by their velocity gives the distance of the round-trip path.

It is not feasible to use radar waves in the classroom, but microwave apparatus has recently become available for classroom use. However, both of these waves are invisible so that to the uninitiated their use more closely resembles hocus-pocus than basic science. In this experiment, we shall use a solid object as an analogy to a radar wave. The success of the experiment depends upon finding an object which will bounce back (reflect) at approximately the same velocity that it was sent out. Fortunately, super-balls meet this requirement. The results of the experiment will not be satisfactory, however, if ordinary rubber balls, steel bearings, or other balls are substituted.

MATERIAL

- Superball
- Ruler (with groove, for launching ramp)
- Books
- Meter stick
- Stopwatch or wall clock

STUDENT PROCEDURE

In this experiment, we want to study how a moving object can be used to measure distances. Astronomers use radar waves as a moving object to measure the distances to the moon and nearby planets. We shall launch a superball from a ramp to measure the distance to a nearby wall, as shown in Figure A-9. Although the final answer will be much more mundane, the principle involved in measuring wall distance and moon distance is the same.

1. Put one end of a ruler on a pile of books to form a launching ramp. We want to measure the velocity of a superball when it is always launched from the same spot on this ruler.
2. Note the time and release the superball. Stop the ball after five seconds and mark its final position.
3. Measure the distance traveled by the superball in five seconds, using the meter stick. Calculate its velocity in m/sec.
4. Repeat steps 1 through 3 until you have decided on a value for the average velocity of the superball.

Now you are ready to measure the distance to the wall. Launch the ball, and measure the time required for the ball to go to the wall and return.

Figure A-9. *A superball rolled from a ruler can be used to measure distances by a method analogous to radar.*

ANALYZING RESULTS

Using the value determined for the average velocity of their super-ball, have students calculate the round-trip distance traveled by the super-ball. From this distance, determine the distance from the launching ramp to the wall.

Students also may wish to measure this distance directly with a meter stick. Ask them to explain any discrepancies between the two values.

We assume that the velocity of the superball is constant. Have students discuss the type of error this introduces into the experiment.

INVESTIGATING FURTHER

The U. S. Army Signal Corps first sent a radio signal to the moon in 1946 and received the echo back in about 2.6 seconds. Radio signals travel through space at the speed of light: approximately 299,800 km/sec. Calculate the distance to the moon from this information.

6

Do They Move or
Do We Move?

EVIDENCE FROM THE FOUCAULT PENDULUM

MOTIVATORS

In previous experiments, students have measured the angular velocities of stars, the sun, and the moon. Subsequent measures of the distance to the moon in Astronomy Experiment 4 led to the calculation of an enormous value for its apparent linear velocity. Some students may have suspected for some time now that most of the apparent angular displacement could be accounted for by the rotation of the earth. In this experiment, we shall consider some proof that the earth is rotating.

BACKGROUND INFORMATION

Many of the apparent motions in the sky can be explained by assuming that the earth rotates on an imaginary axis. It was not until 1851, however, that the first widely accepted proof of the earth's rotation was provided. In that year, the French physicist Foucault performed a now-famous pendulum experiment at the Pantheon in Paris. Foucault hung a heavy iron ball from the dome of the Pantheon by a 200-ft wire, and set it swinging in a north-south direction. Although a freely swinging pendulum should not change its direction, Foucault's pendulum steadily appeared to change direction until, about eight hours later, it was swinging in an east-west line.

That such an experiment should somehow prove the earth's rotation is not apparent to most people until they investigate the behavior of a swinging pendulum supported from a slowly rotating platform. This experiment begins with such an investigation.

MATERIAL

- 2 wire clothes hangers
- Small weight
- Thread
- Plastic 1-gal bottle
- Sand
- Heavy cord or twine

STUDENT PROCEDURE

Perhaps previous experiments have led students to guess that many of the apparent motions of the moon, sun, and stars can be explained by assuming that the earth rotates. Can we do an experiment to show that the earth does rotate? We would have to look for some unusual behavior of a laboratory object which could be explained by that rotation. All the usual objects in the laboratory are carried along with the earth so that relative to its surface they seem to stand still. Perhaps we should look for something which can move freely for a relatively long period of time. A swinging pendulum is a good choice.

How does a pendulum swing when its support rotates? We can easily find out.

1. Bend two wire clothes hangers together so that they form a frame like the one pictured in Figure A-10, next page. Tie a small weight to one end of a thread and tie the other end of the thread to the frame so that the weight can swing like a pendulum. Tape a piece of paper across the bottom of the frame.
2. Start the weight swinging. Draw a line on the paper below the weight to record the direction of the swing. (To increase the students' accuracy, you could hang a light bulb directly over the pendulum and trace the path of the moving shadow of the weight.)
3. Now slowly turn the frame at least 45°. Draw the path of the swing of

223

the pendulum on the paper once again. Ask students to compare this path to the one they drew earlier.

4. Holding the frame at arms length, turn it slowly. Does the pendulum appear to turn? Now place your face close to the frame. Turn the frame, moving your face with it. Does the pendulum appear to turn?

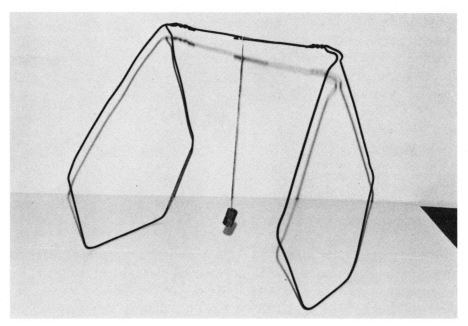

Figure A-10. *Two clothes hangers form a pendulum support which can be rotated.*

ANALYZING RESULTS

Have students measure the angle formed by the two lines of swing which they drew. Ask how this angle compares to the angular displacement of the frame when it was rotated. Ask how the directions of change compare (describing them as clockwise or counterclockwise).

Have students imagine that they were riding on a rotating frame. Ask them to describe how the motion of the pendulum would appear to them. Ask them to describe how they might measure the rate of rotation of the frame by making measurements on the pendulum.

INVESTIGATING FURTHER

It is not difficult to make an actual Foucault pendulum which will show rotation within a class period. The most crucial part of such a pendulum is its support, which must be able to rotate freely. Commercial Foucault pendulum supports are available for about $34; however, a satisfactory one can be devised from an ordinary ice pick and a tin can lid.

Figure A-11 shows how the support is made. The ice pick is bent so that the weight of the pendulum will ultimately rest on the sharp point of the pick. If the ice pick is made from tempered steel, it will be necessary to heat it at the two points where it is to be bent. The point of the ice pick should rest in a small depression made in a tin can lid with a nail or a flat-ended screw. A drop of oil placed in this depression will allow the ice pick to rotate freely.

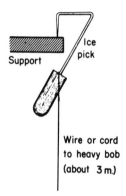

Figure A-11. *Foucault pendulum support.*

The speed of swing of a pendulum does not depend upon the mass of the pendulum bob. However, a heavy bob will possess more inertia so that the swinging motion will not die down as rapidly as with a light bob. A gallon plastic bottle filled with sand will make a good bob. Tie the bob to the handle of the ice pick with a cord which is about 3 m long (or longer). Prepare the pendulum at least two or three hours before class time to allow the cord to unwind slightly and become stable.

Gently set the pendulum swinging. A meter stick can be placed on the floor beneath the pendulum in line with its direction of swing so that later changes can be detected. In about 30 or 40 minutes, students should be able to notice that the path of swing appears to have rotated clockwise (when viewed from the top). The rotation angle in the path depends upon latitude, and will probably be no more than about 10°/hr.

Ask students what the rate of apparent angular rotation would be if the pendulum were supported directly over the earth's axis of rotation? (The time for a complete revolution would then be 24 hours.)

7

Do They Move or
Do We Move?

EVIDENCE FROM SUNSET AND MOONLIGHT

MOTIVATORS

Let's make a picture of the motion of the sun, moon, and earth as we know from experiments so far. We have discovered that most of the apparent motions seen in the sky can be explained by the rotation of the earth. Suppose this spin of the earth is the only actual motion of these three objects. If they were arranged something like Figure A-12, then we should see first the sun and then the moon in the sky as the earth rotated. Of course, if they were all in a straight line as the figure shows, the earth would block off the sun's light so that the moon would always be dark. We can get around this problem by imagining that the earth is pulled out of the plane of the paper just a little bit. Ask students to describe what they would see if the earth, sun, and moon were arranged this way and if only the earth rotated. Would there be any difference in what they saw if they stood first at the equator and then at the North Pole? Would they ever see the sun and the moon in the sky at the same time?

BACKGROUND INFORMATION

Background information for this experiment is included as part of the "Analyzing Results" section.

226

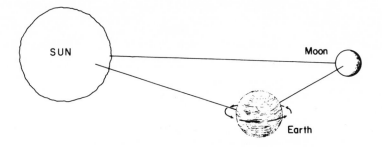

Figure A-12. *Sunrises and moonrises would never vary if all the apparent motions in the sky were caused by the earth's rotation.*

MATERIAL

- Graph paper

STUDENT PROCEDURE

If all the apparent motions observed in the sky were caused by the constantly spinning earth, then these motions should repeat themselves exactly every 24 hours (the period of the spin). That is, the sun should rise and set at the same times every day. If the moon (or a star) is at a certain position in the sky at a given time, it should appear to have returned to that position 24 hours later.

Table A-1 gives the time the sun is on the western horizon (sunset) and the time the moon is at its highest point (meridian passage) for Wednesday, January 1, 1969 and each Wednesday thereafter until Wednesday, December 31, 1969. Have students graph the time of sunset against the number of the Wednesday, and (on the same graph) the time of the moon's meridian crossing against the number of the Wednesday. This will require two standard 8½" x 11" sheets of graph paper taped together to form a large 17" x 11" sheet. Plot the time on the vertical (11") axis beginning with 0 for midnight, continuing through the AM hours, 12 noon, the PM hours, and ending once again with 0 for midnight. A convenient scale is to let the vertical side of each square represent a half hour. Plot the number of the Wednesday on the horizontal axis, letting the horizontal side of each square represent one week.

Table A-1. *Motions of the sun and moon for 1969.*

Week Number	Wednesday	Time of Sunset (PM)	Time of Moon's Meridian Passage
1	*Jan.* 1	4:29	10:46 PM
2	8	4:36	3:37 AM
3	15	4:44	9:25 AM
4	**22**	4:53	4:07 PM
5	29	5:03	9:31 PM
6	*Feb.* 5	5:13	2:19 AM
7	12	5:23	8:15 AM
8	19	5:33	2:42 PM
9	26	5:43	8:15 PM
10	*Mar.* 5	5:52	12:59 AM
11	12	6:01	7:11 AM
12	19	6:10	1:17 PM
13	26	6:19	6:58 PM
14	*Apr.* 2	6:28	
15	9	6:37	6:07 AM
16	16	6:46	11:55 AM
17	23	6:55	5:39 PM
18	30	7:03	10:59 PM
19	*May* 7	7:12	5:01 AM
20	14	7:20	10:36 AM
21	21	7:28	4:22 PM
22	28	7:35	9:36 PM
23	*June* 4	7:41	3:50 AM
24	11	7:46	9:19 AM
25	18	7:49	3:05 PM
26	25	7:51	8:14 PM
27	*July* 2	7:50	2:33 AM
28	9	7:48	8:03 AM
29	16	7:44	1:49 PM
30	23	7:38	6:57 PM
31	30	7:30	1:13 AM
32	*Aug.* 6	7:21	6:47 AM
33	13	7:10	12:31 PM

34	20	6:59	5:45 PM
35	27	6:47	
36	*Sept.* 3	6:35	5:30 AM
37	10	6:22	11:12 AM
38	17	6:08	4:38 PM
39	24	5:55	11:19 PM
40	*Oct.* 1	5:42	4:12 AM
41	8	5:29	9:48 AM
42	15	5:16	3:37 PM
43	22	5:04	9:55 PM
44	29	4:54	2:54 AM
45	*Nov.* 5	4:44	8:23 AM
46	12	4:53	2:25 PM
47	19	4:28	8:41 PM
48	26	4:23	1:38 AM
49	*Dec.* 3	4:20	6:58 AM
50	10	4:18	1:13 PM
51	17	4:20	7:25 PM
52	24	4:23	12:23 AM
53	31	4:28	5:34 AM

ANALYZING RESULTS

When the points plotting the times of sunset are joined by a smooth curve, the result is a sinusoidal-shaped graph. (It is mathematically incorrect to connect these points with a curve, since the sunset time is not a continuous function. However, drawing such a curve will aid in studying the resulting pattern for both the sun and moon data.) Students may also notice that if the left and right edges of the graph are joined by bending the paper to form a vertical cylinder, the end points of the graph curve match (to within one minute). This result suggests that the curve is periodic, repeating itself every year. What yearly motion of the sun could produce this effect?

Students may have some doubt about how to connect the points plotted in connection with the moon timings, since these points are spaced relatively far apart. The important thing to realize is that they must be connected in the same sequence in which they were plotted. Thus the

229

point for the second Wednesday must be connected to the point for the third Wednesday, etc. When the times change from PM to AM the line should be broken off, since this is a result of the way we chose the vertical axis. If these instructions are followed, the result will be a series of steeply rising straight lines.

Like the sun pattern, this pattern repeats itself, but much more frequently. A year is required for the sun pattern to repeat. The time (period) for the moon pattern to repeat can be estimated by noting the number of days represented by the space between the intersections of two consecutive graph lines with the horizontal axis. (Remember that the space between two consecutive Wednesdays represents seven days.) The accuracy of this reading will not be great because of the scale suggested; nevertheless, students should arrive at an estimate close to 30 days. What "monthly" motion of the moon could produce such an effect? Be sure to lead students in a discussion such as the one suggested in the following section.

INVESTIGATING FURTHER

Imagine that the sun revolved in a large circle around the earth, in the same direction as the earth's spin. Because such a sun would move a little bit every 24 hours, it should set a little later every day, since it would take just a little longer for the earth's rotation to catch up to the new position of the sun. In the first part of the year this is exactly what happens. Very observing students may realize that our revolving sun model would predict that catch-up time should be exactly the same each day (or for each seven-day period). This, in turn, should lead to a straight-line graph. But the problem is even worse! After the 26th Wednesday, the sun sets earlier each evening. This would imply that the sun is moving counter to the direction of the earth's revolution. Thus instead of revolving, the sun would have to be moving back and forth like a lethargic cosmic pendulum. Can we form an alternative model?

We can observe that together with changing sunset times, the sun's position changes as it sets on the horizon. This position moves north during the spring and summer and south during the fall and winter. Furthermore, the star constellations move across the sky with a yearly regularity. These two observations suggest that the earth may have a yearly revolution as well as a daily rotation. The swinging effect of the sun in the sky can be explained by assuming that the earth's imaginary axis of rotation, while

230

slanted with respect to the plane of the revolution, remains always parallel to any given previous position throughout the year. You may want to have students demonstrate this model using small rubber balls and flashlights. This is also an appropriate time to study star maps.

Although the "revolving sun model" suggested in the above paragraphs must be discarded because of complexities, the same is not true for a revolving moon model. Inspection of the graph shows that the moon does continue to return to a given position at a later time each day, and in a constant or linear fashion of about six hours each week or 50 minutes each day. Final confirmation of the revolving moon model lies in its ability to explain moon phases.

8

How Big Is the Moon?

MOTIVATORS

Previous experiments gave students information about the distances from the earth to the moon, and about motions of the moon and the sun. In this experiment, students will investigate and measure the diameter of the moon (and optionally, the sun).

BACKGROUND INFORMATION

Hold your thumb up at arm's length in front of you. Imagine lines drawn from either side of your thumb to a point on the pupil of your eye. The angle formed by these two lines is a kind of measure of the width (diameter) of your thumb. This angle is called the *angular diameter* (of your thumb, in this case). Now move your thumb closer to your eye. What happens to the size of the angular diameter? The value of an object's angular diameter depends upon the distance from the observer to the object. We can use this same idea to measure the diameter of the moon, since we know the distance from the earth to the moon. Figure A-13 shows the lines we will consider.

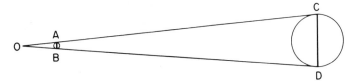

Figure A-13. *The length of CD can be calculated if one knows the length of AB and the distances OB and OD.*

Suppose one holds a pencil in front of his eye (point 0 in Figure A-13) until it just appears to cover the moon. The distances AB (the diameter of the pencil) and OA (the distance from the pencil to the eye) can be easily measured. The distance OC is known to be approximately 385,000 km. We could essentially construct a scale drawing to measure the distance CD (the diameter of the moon).

In practice, the triangle OAB is far too small in relationship to triangle OCD to be reproduced accurately on a scale drawing. However, since triangles OAB and OCD are similar, the ratio of sides CD to AB is the same as the ratio of sides OC to OA. That is,

$$\frac{CD}{AB} = \frac{OC}{OA} .$$

You may wish to have younger students draw triangles similar to Figure A-13, measure the two pairs of corresponding sides, and compare their ratios to establish this principle.

Since the moon may not be available to you during your class period, you may want to use a 40-watt light bulb on your desk as an artificial moon. Students will need to know the distance from their observation positions to the light bulb. In the laboratory this can be measured directly, of course. After students have measured the diameter of the light bulb indirectly, they can measure the diameter of the moon at home.

MATERIAL

- Meter stick
- Pencil (preferably with a pocket clip)
- 2 pieces of cardboard (approximately 12 x 15 cm)
- Thumbtacks
- 40-watt light bulb and socket

STUDENT PROCEDURE

The diameter of the moon can be measured by adhering to the following procedure:

1. Make a pinhole in a piece of cardboard (about 12 x 15 cm). Thumbtack the cardboard on the zero end of a meter stick so that you can see along the length of the stick when you look through the pinhole.
2. Point the stick at the moon (or an object at the front of the classroom), and sight through the pinhole at the moon (or object).
3. Slide a pencil back and forth along the meter stick until the thickness of the pencil just covers the width of the moon (or object). If the pencil has a pocket clip, fasten it onto the meter stick at this position.
4. Read the distance from the pinhole to the pencil. Find the ratio of the distance from the pinhole to the pencil to the distance from the pinhole to the moon (or object). (The distance from the earth to the moon is approximately 385,000 km. If you are using an object at the front of the classroom, you will have to find the distance by measuring it directly.)

ANALYZING RESULTS

The value for the ratio found in step 4 of the Student Procedure section must be the same as the value for the ratio of the diameter of the moon to the diameter of the pencil. (See Figure A-13.) Measure the diameter of the pencil and calculate what the diameter of the moon must be.

The radius of the earth is approximately 6400 km. What is the distance from the center of the earth to the center of the moon?

INVESTIGATING FURTHER

A variation of this same method can be used to measure the diameter of the sun. The sun cannot be sighted directly without risking damage to the eyes. However, the sun is bright enough to shine through the pinhole and cast its image on a second piece of cardboard held behind the pinhole, as shown in Figure A-14. The diameter of this image, AB, can be marked on the cardboard so that it can be measured later with a ruler. The distance OA can be assumed to be the same as the distance from the image card to the pinhole card measured along the meter stick. The distance to the sun, OC, is 150 million km.

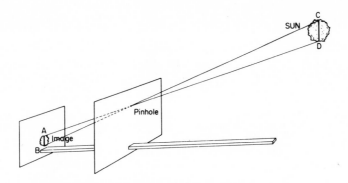

Figure A-14. *The diameter of the sun can be measured from its image cast on a card.*

9

Measuring Distances in
the Solar System

MOTIVATORS

The measurements and constructions which the student will make in this experiment depend upon some general understanding of apparent vs. actual motions. You can start a classroom discussion of relative motion by performing the following demonstration. Make a "V" out of some coat hanger wire and stick one styrofoam ball on each end of the "V." Arrange the wire and balls with a light bulb inside a cardboard box so that the students can only observe the shadows of the styrofoam balls cast on the classroom wall by the light bulb. Holding the wire "V" so that one leg is vertical, rotate the wire around this vertical axis so that one ball moves around the other. What do the shadows on the wall look like? One should appear to stand still (or nearly so), and the other one should move back and forth. Have students guess what kind of motion is actually happening inside the box. Could the shadows be caused by a swinging pendulum? If the motion is circular, what is the size (radius) of the circle?

BACKGROUND INFORMATION

When the planets Mercury and Venus are observed over long periods of time, their motions in the sky look much like the shadows cast on the wall in the preceding experiment. They appear to wander back and forth among the stars, but we cannot be sure whether this is their actual motion or only their apparent motion. We are led to guess circular motion because of the successes of this model in explaining earth motions in the preceding experiment. (If we had a telescope to make careful measurements, we would also observe that the sizes of Mercury and Venus appear to change as they move. Students may have noticed this same effect with the styrofoam ball shadows.)

This experiment investigates the measurement of such observations, and allows students to construct from these observations a scale model of the solar system orbits.

MATERIAL

- Ruler

- Drawing compass

- Protractor, plastic straw, and straight pin assembled as shown in Figure A-5, Astronomy Experiment 2.

STUDENT PROCEDURE

By this time, students should realize that the nature of apparent motion depends upon the nature of the point of observation. Begin this experiment by setting up a planetary model using students from the class. Choose one student to represent the sun and another to represent a planet. Ask the planet student to walk in a circle around the sun student. Have the remainder of the class (sitting at their desks) outstretch both of their arms, pointing the left hand at the person who represents the sun and the right hand at the person who represents the planet. The observers should move their right arms so that they always point at the moving planet student. Ask what kind of motion the right arms make. Discuss how this back-and-forth motion might be measured.

One way of measuring the amount of motion is to measure the maximum angle made between the right and left arms (that is, the largest angle between the lines of observation to the sun and the planets). Can this maximum angle be used to measure the size of the planet's orbit? The answer is yes, provided the distance from the observer on the earth to the sun is known. The following steps illustrate the required method, using the protractor-straw measuring device from Astronomy Experiment 2.

237

1. Choose an observer, planet, and sun from the class. Let the sun student and the observer stand about 4 m apart. Measure and record this distance.
2. Now ask the planet student to walk slowly around the sun student, in a circle whose radius is about 2 m.
3. The observer can now use the protractor-straw device to measure the maximum angle between the observed positions of the sun and the planet. Holding the protractor steady, point the straw at the sun and read the angle. Then point the straw at the maximum outward position of the planet and again read the angle. The difference between the two angle readings is the maximum angle between the sun and planet.
4. Make a scale drawing of the situation just observed. Choose a scale (5 cm = 1 m is a good choice if students need specific suggestions). Draw a straight line whose length represents the distance from the observer to the sun, according to the chosen scale. Label one end of the line "sun" and the other end "earth." From the earth point draw the line of observation to the apparent maximum displacement of the planet. The angle between the earth-sun line and the earth-planet line will be the same as the maximum angle measured by the observer. (The drawing should resemble the sketch in Figure A-15.)

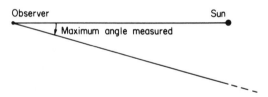

Figure A-15. *If the maximum angle between an inner planet and the sun is known, that planet's orbit can be constructed to scale.*

5. Using a compass, draw a circle whose center is the sun and which just touches the earth-planet line. Measure the radius of this circle, and convert the value to actual distance using the scale factor. How does this value compare to the distance between the "sun student" and the "planet student?"

ANALYZING RESULTS

More than 300 years ago an astronomer named Tycho Brahe made measurements of the maximum angles between the sun and the planets Mercury and Venus. Tycho Brahe did not know the value for the distance between the earth and the sun. However, if this distance is called one unit (astronomers refer to it as one astronomical unit, or 1 a.u.), a scale drawing can be made which will give values for the radii of the Mercury and Venus orbits in fractions of this unit. Let 5 cm represent 1 a.u. Construct a scale drawing of these planetary orbits if the maximum observed angle between the sun and Mercury is 24°, and if the maximum angle between the sun and Venus is 48° (using the instructions given in steps 4 and 5 of the student procedure). Find the values for the radii of these orbits in terms of astronomical units.

The method we have just described works well for finding the radii of the orbits of planets between the earth and the sun. But what about planets outside the earth's orbit? The following steps for constructing the orbit for Mars illustrates how they are found.

Letting 5 cm = 1 a.u., construct a circle representing the earth's orbit. It takes 365 days for the earth to travel once around this orbit. But Mars takes 687 days to go around the sun. That is, while Mars makes one trip around the sun, the Earth goes around 687/365 or 1.88 times. Place a dot on the earth orbit to represent its position today. Knowing that 1.88 revolutions = 677°, measure and mark the position of a second dot which will represent the position of the earth 687 days later.

Tycho Brahe had measured the angle between the sun and Mars to be 119° on a certain day. Construct this line of observation on the scale drawing (the 119° angle should be constructed so that the second earth dot is in the interior of the angle). Six hundred eighty-seven days later Tycho measured the angle between the sun and Mars to be 136°. Construct this second observation line, using the second earth position. Remember that Mars is in the same position for both observations (see Figure A-16). Where do the lines intersect? How far, in astronomical units, is Mars from the sun?

Tycho Brahe also made observations for the planets Jupiter and Saturn. Using similar constructions, he found Jupiter to be 5.2 a.u. from the sun and Saturn to be 9.5 a.u. from the sun. He did not suspect any more planets. We now know that there are three more—Uranus at a dis-

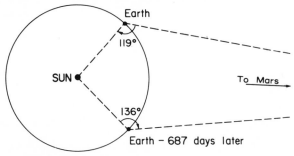

(Not drawn to scale)

Figure A-16. *A scale drawing for determining the orbit of Mars should be arranged like this sketch.*

tance of 19 a.u., Neptune at 30 a.u., and Pluto at 39.5 a.u. from the sun.

INVESTIGATING FURTHER

How big are these astronomical distances in kilometers? If we knew the actual size of any one distance in kilometers we could figure them all out. The distance from the earth to Venus can be measured by sighting from two points and making a scale drawing of the resulting triangle, just as students did in finding the distance to the moon. But since Venus is much farther away, the lines are very close to being parallel, and the drawing requires much more accurate instruments than students have available. In 1961, radar waves were bounced off the surface of Venus. The round trip took 140 seconds, which gives approximately 42 million km as the distance from the earth to Venus. This information is sufficient for converting astronomical units to kilometers.

Information given here will allow students to construct a scale drawing of planetary positions in the solar system but says nothing about the sizes of the planets in relationship to this scale. Table A-2 is a workable scale which includes planetary diameters as well as distances. You may wish to have students construct a scale model of the solar system using these values. (Drawing a dot whose diameter is only 0.05 mm may cause considerable consternation! Some students may prefer to multiply everything in the table by a factor of 10 or more, and this is permissible if they can find paper big enough.)

Table A-2. *Solar-system scale.*

Object	Distance from Sun (m)	Diameter (mm)
Sun	0	11.00
Mercury	0.4	0.04
Venus	0.7	0.10
Earth	1.0	0.10
Mars	1.5	0.05
Jupiter	5.2	1.12
Saturn	9.5	0.95
Uranus	19.2	0.37
Neptune	30.1	0.35
Pluto	39.5	0.10

10

Celestial Mechanics

MOTIVATORS

$\mathbf{I}$n previous experiments, we have explained the changing positions of objects in the solar system by imagining that they are moving in circles. Can an object move in a circle all by itself? Have students whirl a heavy object, like a bucket of sand, around in a circle. They should realize immediately that circular motion requires a considerable force! What is the nature of this force? Where does it come from in the solar system? In this experiment, we shall attempt to discover answers to these questions.

BACKGROUND INFORMATION

Background information for this experiment is included as part of the "Analyzing Results" section.

MATERIAL

- Rubber stopper

- String, tape

- Metal washers

- Paper clips

- 6″ long glass tube

- Stopwatch, or watch with a sweep second hand

STUDENT PROCEDURE

We can now be fairly sure that planets move in circles around the sun. The moon moves in a circle around the earth. What kind of forces are necessary for circular motion? Students can investigate this question using the very simple apparatus shown in Figure, A-17. A 45-cm length of string, threaded through a 15-cm length of glass tubing, is tied to a rubber stopper at one end and to a paper clip at the other end. (The ends of the glass tube should be smoothly fire-polished, and the tube itself wrapped with tape to prevent shattering if it should accidentally be broken.)

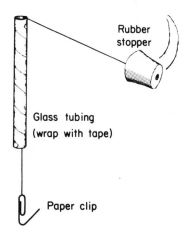

Figure A-17. *Centripetal force can be investigated using this apparatus.*

Bend the wire end of the paper clip to form a hanger for several metal washers. Let students investigate the requirements necessary for whirling the rubber stopper around in a circle, using the tube as a handle. They should see quickly that the washers must pull on the stopper, and that the faster the stopper whirls, the larger this pull must become. Use the following steps to discover the relationship between the speed of the whirling washer and the force required.

1. Place five washers on the paper clip hanger. Whirl the rubber stopper until the paper clip moves to just below the bottom of the glass tube. (Be sure that the only force pulling on the stopper is supplied by the weight of the five washers!) Have a second student measure the time required for 10 revolutions, and calculate the time for one revolution. Record the weight (number of washers) and period (time for one revolution).

2. Repeat step one using 10, 15, 20, and 25 washers, in turn. For each trial record the weight and the period.

3. Graph the force, F (vertical axis), against the period, T (horizontal axis). Mathematicians know that the equation of a straight-line graph which passes through the origin is of the form $y = kx$. The F vs. T graph should be a curve. Can we obtain a straight-line graph from our data by making proper adjustments? Have students calculate $1/T^2$ for each trial. Graph force, F, vs. $1/T^2$.

ANALYZING RESULTS

Depending upon the accuracy of each trial, the F vs. $1/T^2$ graph should be approximately straight, and should pass close to the origin. This means that the mathematical relationship between the necessary inward, or centripetal force, for circular motion whose period is T is given by the equation

$$F = k(1/T^2).$$

Further experiments have shown the value of the constant k to be $m4\pi^2R$ where m is the mass of the stopper, and R is the radius of the circle. That is, all circular motion requires a centrally directed force whose size is given by the equation

$$F = \frac{m4\pi^2R}{T^2}.$$

Most students probably have heard of Newton's law relating force and the acceleration which it produces. This relationship is given by $F = ma$. If the law is applicable to circular motion, then it implies that when an object moves in a circle it continually accelerates toward the center of the circle by an amount

$$a = \frac{4\pi^2R}{T^2}.$$

243

If you imagine an object moving in a circle, you can explain a piece of its path as a straight line which has been bent toward the center of the circle. The rate at which this bending of its straight-line velocity occurs is the acceleration given by the preceding formula.

INVESTIGATING FURTHER

We now have enough information to calculate the acceleration of each planet as it moves around the sun. Table A-3 gives the average radius for the planets in meters, and the time for one revolution (period) in seconds.

Table A-3. *Orbited radii and periods for the planets.*

Planet	Radius, R	Period, T
Mercury	0.58×10^{11}	0.76×10^{7}
Venus	1.1×10^{11}	1.9×10^{7}
Earth	1.5×10^{11}	3.2×10^{7}
Mars	2.3×10^{11}	5.9×10^{7}
Jupiter	7.8×10^{11}	37.9×10^{7}
Saturn	14.3×10^{11}	93.0×10^{7}
Uranus	28.7×10^{11}	266.0×10^{7}
Neptune	45.0×10^{11}	520.0×10^{7}
Pluto	59.0×10^{11}	782.0×10^{7}

Using the formula $a = 4\pi^2 R / T^2$, students should be able to calculate values for the acceleration of each planet. They should obtain a table much like Table A-4.

Table A-4. *Centripetal acceleration of the planets.*

Planet	Acceleration (m/sec^2)
Mercury	$39{,}000 \times 10^{-6}$
Venus	$12{,}000 \times 10^{-6}$
Earth	6300×10^{-6}
Mars	2700×10^{-6}
Jupiter	210×10^{-6}
Saturn	64×10^{-6}
Uranus	16×10^{-6}

Neptune	6.5 x 10^{-6}
Pluto	4.3 x 10^{-6}

What can we tell about these accelerations? One of the most striking features is, of course, that the acceleration drops rapidly as the distance from the sun (radius of the orbit) increases. Notice that Uranus is about twice as far from the sun as Saturn; the acceleration of Uranus is ¼ that of Saturn. Does this mean that the relationship between the acceleration toward the sun and the distance from the sun is of the form

$$a = k \ (1/r^2) \ ?$$

Students can check by graphing acceleration, a , against $1/r^2$. A graph of these variables for the last four planets may be enough to check this hypothesis. The results should be a straight line passing very close to the origin, confirming our guess.

We will conclude by checking the acceleration of the moon around the earth. The distance from the center of the earth to the moon is 3.8 x 10^8 m, and the period of the moon's revolution is 2.3 x 10^6 seconds. Using the formula

$$a = 4\pi R/T^2,$$

students should obtain a value of 2.7 x 10^{-3} m/sec^2. Now let's suppose that this acceleration is caused by the earth. If you drop a ball at the surface of the earth, its acceleration towards the center of the earth is 9.8 m/sec^2. The radius of the earth is 6.38 x 10^6 m. The moon is about 60 times as far away from the center of the earth as the ball is when it is dropped at the surface of the earth. Then the acceleration on the moon should be $(1/60)^2$ that of the acceleration on the ball. That is,

$$a_{moon} = (1/60)^2 \ (9.8 \ m/sec^2) = 2.7 \ x \ 10^{-3} \ m/sec^2,$$

which is a remarkable agreement!

What does all this mean? All the celestial objects in our solar system accelerate because they move in orbits which are nearly circular. This acceleration is toward the center of that circle, whether the center is the sun or the earth (in the case of the moon). Its value varies as the reciprocal of the square of the distance separating the moving body from the body in the center. Similar statements can be made about the force causing this acceleration, a force we call gravitational attraction. The relationships, or laws, we have investigated keep our solar system in motion like a gigantic celestial clock.

11

Measuring Distances
Beyond the Solar System

MOTIVATORS

$\mathbf{W}$e have spent much time measuring distances within the solar system. How can we measure distances outside the solar system? The method of triangulation used to measure the distance from the earth to the moon becomes very difficult with distant objects, since the two long sides of the triangle drawn out from the base line become more and more nearly parallel. One way to improve this method is to increase the length of the base line. We can do this by taking two observations exactly one-half year apart. The base line between these observations is then twice the radius of the earth's orbit, or 300 million km. But even with such an enormous base line, the apparent shift of the stars is extremely small. It was not until 1838 that instruments were developed which were accurate enough to measure a difference in angle between two such observations. The difference in angle for the star 61 Cygni is 0.67 seconds (where 1 sec = $1/360°$).

BACKGROUND INFORMATION

Alpha Centauri is the nearest star found so far. Its parallax angle (half the angular distance between the observations) is only 0.8 second. Its distance is 42,000,000,000,000 (42 million million) km away, or about 4½ light years (the distance light travels in one year).

Up to now astronomers have measured the parallaxes for about 10,000 stars. After this, distances must be measured by considering the intensity of the light from the stars which reaches us. Using the light meter calibrated according to Appendix IV, we will investigate some of the problems involved in making such measurements.

MATERIAL

- Light meter (calibrated according to Appendix IV)
- Meter stick
- 40-, 60-, and 100-watt light bulbs
- Light bulb socket and cord
- Aluminum foil

STUDENT PROCEDURE

Can we measure the distance to stars by measuring how brightly or faintly they shine? The following steps will help answer this question:

1. Set up a 40-watt light bulb. Make a box out of aluminum foil and set it around (but not touching) the light bulb. Use a pencil point to make a hole (about 8 mm in diameter) in the box at the level of the bulb. The light bulb will be a "star" and the box will prevent extra light from the star from bothering other groups.

2. Before turning on the "star," measure the background light in the room by reading the light meter. Subtract this value from all the readings taken in steps 3 and 4.

3. Place the light meter about 10 cm from the star. The light coming from the star is scattered from the frosted glass, which acts as if it were the source. Make the distance measurement from the glass rather than from the box. (If you are using clear bulbs, the measurement should be made from the filament in the center of the bulb.) If the needle on the light meter goes off scale you can increase this 10-cm distance and all the distances given subsequently by appropriate amounts. Record the reading of the light meter and the distance from the bulb to the meter.

4. Measure the intensity of the light striking the light meter by recording

247

its readings at distances of 20, 30, 40, 50, 60, 70, and 80 cm. Be sure that the meter is at the same height as the hole in the aluminum box each time.

5. Graph the intensity, I (light meter reading), against the distance, D.
6. Set up another 40-watt bulb with an aluminum box on a demonstration desk. Ask students to bring their light meters and measure the intensity at some unknown distance. Have them find a value for this distance using the graph made in step 5.

ANALYZING RESULTS

Try replotting the data to see if a straight-line graph can be obtained. Students may want to try plotting I vs. $1/D$ and I vs. $1/D^2$.

INVESTIGATING FURTHER

The astronomer compares the light intensity of different stars with that of stars that are close enough to be measured by triangulation. But stars are not 40-watt light bulbs. Is it possible that some stars give off more light than others; that they have different luminosities?

Have students repeat the experiment for 60-watt or 100-watt light bulbs. Compare the resulting graphs (an overhead projector is excellent for this). How could one tell which graph line to use in measuring the distance to an unknown star? In the next experiment, we shall investigate this question more thoroughly.

12

Messages from Stars

We have seen that starlight can give us in-
formation about the distances to stars provided we know that we are
comparing two stars with the same luminosity. In this experiment, we shall
see that starlight can be broken down to give us answers about this
luminosity and about chemical composition as well.

BACKGROUND INFORMATION

Much can be learned about stars by studying the *spectrum* (component
colors) of the light they emit. Atoms emit precise colors of light when
electrons move from an outer orbit to an inner orbit. Since the total
numbers and maximum numbers of electrons in each orbital shell are
different for the atoms of each different element, the exact colors present
provide us with clues to the chemical composition of the stars.

Atoms can also absorb light energy, but only in the precise amount
necessary to raise an electron from its orbit to one of the outer orbits.
Such absorption is responsible for a dark line at the appropriate wave-
length in the spectrum of the light. Thus both lines which are present

249

and lines which are absent are important in studying stellar spectra.

In addition, atoms of gases in the atmosphere of a star may lose one or more electrons completely from energy imparted by collisions with other atoms. Such electron-deficient atoms are called *ions*. Their presence depends upon the frequency of such collisions, and these in turn are a function of the star's temperature. Hotter stars emit more lines in the blue end of the spectrum; cooler stars emit more lines in the red end of the spectrum. By matching the spectrum of an unknown star to that of a star whose brightness and distance is known, the luminosity and distance of the unknown star can be computed.

Finally the spectra of stars can give information about whether the star is approaching us or receding from us. This motion causes a shift of the total spectrum of the star; a phenomenon known as the *Doppler effect*. If a star is approaching us, all lines will be displaced slightly toward the blue end of the spectrum. If the star is receding, the lines are shifted towards the red end of the spectrum. The velocity of the star can be calculated from the amount of the shift.

A basic understanding of this information from starlight requires a thorough background in atomic and molecular physics. The purpose of this experiment is only to investigate the existence of these phenomena in an elementary way.

MATERIAL

- Hand-held spectroscopes
- 25-, 40-, 60-, 100-watt light bulbs
- Aluminum foil
- Bunsen burner
- Nichrome wire
- Calcium chloride, copper chloride, potassium chloride, sodium chloride
- Series-wired outlet box (described below) or variable transformer

STUDENT PROCEDURE

We want to investigate light from sources of different luminosity. It is possible to dim a light bulb either by connecting it to a variable transformer or by wiring another light bulb in series with it. A special extension cord which will connect two light bulbs in series can be made

250

in the following way. Most electrical sockets are wired as shown in Figure A-18. However, if you break the connection between the two screws on one side, and connect the wires as shown in Figure A-19, the circuit will be complete only when bulbs are in both sockets. The electricity must flow through both bulbs and the luminosity will be reduced.

Figure A-18. *A parallel-wired socket has one wire connected to each side of the receptacle. Note the metal tab below the two screws.*

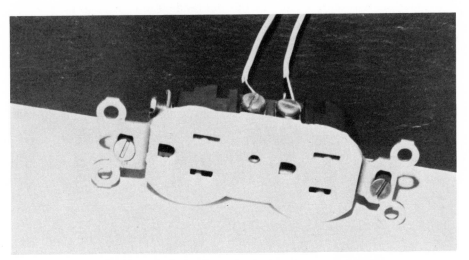

Figure A-19. *A series-wired socket has both wires connected to the same side of the receptacle. Note that the metal tab below the two screws has been broken off on this side.*

1. Place a 100-watt light bulb and a 25-watt bulb in the series outlet. Cover the 25-watt bulb with an aluminum foil box. Look through the spectroscope at the light from the 100-watt bulb.
2. Now look at a 100-watt bulb in an ordinary socket using the spectroscope. Describe the difference in the colors you see when observing first the bright 100-watt bulb and then the dim 100-watt bulb.
3. Look at the spectra of light given off by the 100-watt bulb when it is connected in series to first a 40-watt bulb and then a 60-watt bulb. Do you observe anything that would indicate that the brightness or temperature of a light source has an effect on the colors of light given off?

ANALYZING RESULTS

Students should note that the cooler the star, the "redder" the light. As the luminosity of the source decreases the blue colors disappear from the spectrum. Astronomers check to see that two stars they are comparing have the same luminosity by first comparing the spectra of the light emitted by each.

INVESTIGATING FURTHER

Does the nature of the source of light make a difference in the spectra observed? Suppose the source of light is a flame. Use the spectroscope to look at the flame of a Bunsen burner.

Form a small loop in a long piece of nichrome wire. Place a small crystal of calcium chloride, sodium chloride (table salt), potassium chloride, or copper chloride in the loop, in turn, and hold it in the Bunsen burner flame. Describe and record the spectrum given off in each case.

Appendices

Appendix I

PROCEDURES FOR MAKING FOSSILS

A. Method 1

Equipment:

- Fossils selected for reproduction
- Small glass plate
- Liquid latex (also sold commercially as Liquid Rubber)
- Small brush
- Plaster of Paris

In a well-ventilated room, apply three or four coats of liquid latex to the fossils which have been placed on the glass plate. Be sure to coat all surfaces of the fossils. Allow each coat to dry (approximately 20 to 30 minutes is required) before applying the next coat. Allow at least one hour for final drying. When dry, slit the rubber coat (or mold) with a sharp knife or razor blade. Then peel the mold back and remove the fossil. To prevent the inside walls of the freshly prepared mold from sticking together, peel away the mold from the fossil under water. The mold is now ready to be filled with plaster of Paris to make a fossil cast. Pour plaster of Paris into the mold; shake and stretch the mold so that the plaster of Paris fills in every detail in the mold and air bubbles are removed. Allow cast to harden. Then peel rubber mold off and the cast is completed. With a little practice you will be able to make casts of the more intricate fossils with excellent results (Figure 1).

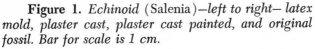

Figure 1. *Echinoid (Salenia)—left to right— latex mold, plaster cast, plaster cast painted, and original fossil. Bar for scale is 1 cm.*

255

Take care when applying the liquid latex on the fossils to get a uniform thickness over all protuberances (such as spine nodes, plates, etc.). In the completed mold, the ends of these protuberances should have small openings to allow the release of air when the mold is filled. This allows complete reproduction of these surface irregularities without entrapping air bubbles. The small openings can be made by snipping off the ends with a fingernail clipper or scissors. Note the excellent detail preserved on the fossil casts. The latex mold can be reused indefinitely.

Adding a coloring agent when mixing the plaster of Paris helps to bring out the detail in the casts and makes the reproduced fossils look more authentic. Materials which have been used with success include water coloring, dye (food coloring), iron oxide, manganese oxide, and powdered paints of various colors. The casts can be painted to achieve this purpose, although the paint may chip off and thus is less effective.

B. Method 2

Equipment:

- Fossils selected for reproduction
- Paper plate
- Plaster of Paris
- Vaseline

Place about ⅓ cup of plaster of Paris on a paper plate. Add water, a little at a time, stirring, until the result is somewhat like thick pancake batter. If too much water is added, the mixture can be thickened by sprinkling on a little more plaster of Paris.

Next grease the fossils to be used with Vaseline. If they have small details such as grooves or pits, be sure that the Vaseline is spread thinly so as not to fill these details. Lightly press the fossils into the plaster of Paris mixture. (Be sure that the top of the fossil is uncovered so that it can be pried out later. If the plaster of Paris mixture is too thin, a heavy fossil may sink. If this happens, pull it out immediately and rework the mixture.)

Set the plate aside. The plaster of Paris should dry in 20 to 30 minutes. (Use the intervening time to discuss prints, molds, and the other kinds of fossil preservation.)

Once molds have been made, it is relatively easy to make

casts. Use Vaseline to grease the mold thoroughly. Too much Vaseline will fill up small grooves, and the resulting cast will lose much of the detail of the original fossil. Make a new plaster of Paris mixture which is a little thinner than the one used previously. Pour this new mixture into each of the molds, and set aside to dry. Remove the casts when they are dry. How do they resemble the original fossils? How are they different? As in the previous method of making fossils, a minor amount of coloring material can be added to the wetted plaster of Paris so as to make the fossils—both molds and casts—appear more realistic and bring out the detail.

Appendix II

PHYSICAL PROPERTIES OF MINERALS

All earth science teachers encounter the problem of identifying minerals and rocks brought to them by students. These specimens are generally common types, and usually found locally. On occasion, a rare specimen is included (perhaps a mineral given to the student by a friend or relative who obtained it while traveling). Such "displaced" specimens may be quite difficult to identify.

Because over 2000 different minerals are known to exist, no "magic formula" can be used to identify all unknown specimens. However, examination of a series of physical properties will aid in the identification of most minerals and confirm the identity of many of the most common ones. This appendix includes a brief summary of the most readily tested physical properties of minerals along with a chart (Table 1, page 261) listing these properties for some common minerals. By examining unknown specimens for these physical properties, students are able to develop a proficiency in mineral identification.

A *mineral* is a naturally occurring, inorganic substance which is characterized by definite physical properties and a chemical composition expressible by a formula. Minerals are made up of atoms and molecules of the various chemical elements and they are the main constituents of all rocks. Physical properties reflect the nature (size, shape, and arrangement) of the atoms and molecules that make up a mineral. For example, the atomic lattice structure controls the cleavage of a mineral, and the density of packing affects the specific gravity and hardness. Thus, the nature of the atomic arrangement as well as the composition of the atoms control the behavior of the mineral under different geological conditions.

Those minerals which are rare or for some other reason are difficult to identify by simple examination of their physical properties, may require more extensive testing. Mineralogists use expensive, highly sensitive instruments (microscopes, X-ray machines, etc.) to gain additional information about these minerals.

The physical properties which can readily be determined

258

for most minerals without special equipment are: (A) luster, (B) hardness, (C) color, (D) streak, (E) transparency, (F) crystal form, (G) break (cleavage and fracture), (H) specific gravity, and (I) other special characteristics such as feel, magnetism, odor, and taste.

Follow the order given below, because by starting with luster, all minerals can be divided into two major groups—metallic and nonmetallic. Further subdivision based on hardness brings the student close to a correct choice of the identity of the mineral being examined.

(A) Luster—The appearance of the mineral surface in reflected light. Minerals can be divided into two major groups on the basis of their luster. One group has a metallic luster (looks like a metal) and is opaque. The other group has a nonmetallic luster and may be either transparent or opaque. Nonmetallic lusters are given special names such as: adamantine (like diamond), dull or earthy, glassy (like broken glass), greasy, pearly, resinous, silky, and waxy.

(B) Hardness—The resistance of a mineral to being scratched. The relative hardness of a mineral can be determined by comparing the specimen with a series of minerals chosen as a relative hardness scale. The standard scale of hardness from 1 (softest) to 10 (hardest) is:

1.	Talc	6.	Orthoclase
2.	Gypsum	7.	Quartz
3.	Calcite	8.	Topaz
4.	Fluorite	9.	Corundum
5.	Apatite	10.	Diamond

In addition to the hardness scale, other common objects used for testing hardness are: fingernail as 2.5, penny as 3, knife blade as 5.5, and window glass as 5.5 to 6.

(C) Color—Although it is the most obvious physical property, it is commonly of limited use because many mineral species have a variety of colors. For example, the mineral quartz is commonly milky white, clear, pink to rose, dark cloudy (smoky), or tints of purple in color. It may also be shades of red, brown, and even jet black among other colors. However, certain broad color groups do prevail for most minerals.

(D) Streak—The color of the powder of a mineral. Commonly, streak is obtained by rubbing the specimen on a ceramic plate or by grinding up a small piece of the specimen. The color of the streak is commonly constant, even though the color of the mineral is variable.

(E) Transparency—The degree to which a mineral will transmit light. A *transparent* mineral is clear, allowing passage of light like windowglass, and images may be observed through a section of the mineral. A *translucent* mineral allows passage of light, but the mineral is not clear and images cannot be observed through a section of the mineral. *Opaque* minerals have no transparency.

(F) Crystal Form—The external shape of a mineral which reflects the internal arrangement of the atoms. Most minerals are crystalline, the regular and definite pattern of the internal atomic structure results in a definite external shape called a *crystal*. A few minerals are amorphous (noncrystalline).

Based on shape and symmetry, all crystals may be systematically placed into one of the following six crystal systems:

1. *Isometric system.* (Three crystal axes of equal length at right angles to each other.)
2. *Tetragonal system.* (Three crystal axes which are at right angles to each other; two taken as horizontal are of equal length, the third is vertical and longer than the other two.)
3. *Hexagonal system.* (Four crystal axes, three of which are horizontal, equal in length, and lie in the same plane making angles of 60° and 120° with each other. The fourth is either shorter or longer than the other three, and is vertical and at right angles to them.)
4. *Orthorhombic system.* (Three crystal axes of unequal length at right angles to each other.)
5. *Monoclinic system.* (Three crystal axes of unequal length. Two of the axes intersect at right angles; the plane formed by these two axes is oblique to the third axis.)
6. *Triclinic system.* (Three crystal axes of unequal length, oblique to each other.)

(G) Break—Cleavage and fracture of a mineral. *Cleavage* is a breaking or splitting that tends to be parallel to crystal faces and yields smooth, flat surfaces along certain directions. Minerals may have one, two, three, or more directions of cleavage. The number, direction, and angle between cleavage surfaces is constant for each mineral.

Fracture is any break other than cleavage. Fracture is described by terms such as conchoidal (like the break of glass), jagged or hackly, splintery, and uneven. Some minerals have both cleavage and fracture; other minerals have only cleavage, and still others only fracture.

Mineral Name	Luster	Hardness	Color	Streak	Transparency	Crystal Form	Cleavage	Fracture	Specific Gravity
Biotite Mica	glassy to pearly	2.0-2.5	dark brown to black	grayish white	transparent to translucent	monoclinic	1 prominent direction (thin sheets)	uneven	2.7-3.1
Calcite	glassy or dull	3.0	colorless, white, or lightly tinted	none	transparent to translucent	hexagonal	3 directions (rhombic)	none	2.6-2.8
Feldspar (variety microcline)	glassy to pearly	6.0	white, pink, gray, or red	white	translucent if thin, otherwise opaque	monoclinic	2 directions at 90° angles	uneven	2.4-2.7
Halite	glassy to dull	2.0-3.0	colorless, gray, or white	white	transparent	isometric	3 directions at 90° angles (cubic)	uneven	2.0
Quartz	glassy	7.0	usually colorless, but can be any color	white	transparent to translucent	hexagonal	none	conchoidal	2.6
Pyrite	metallic	6.0	brassy yellow	greenish black	opaque	isometric	none	uneven	4.9-6.0

Table 1. *Physical properties of some common minerals.*

(H) Specific Gravity—The weight of a mineral compared to the weight of an equal volume of water (relative density). A common method of measuring specific gravity is to weigh a mineral suspended in air and then to weigh it suspended in water. The weight suspended in water is always the weight of the mineral in air minus the weight of water equal to the volume of the specimen. Calculation of specific gravity is as follows:

Let X equal the weight of the mineral specimen suspended in air.

Let Y equal the weight of the mineral specimen suspended in water.

$$\text{Specific Gravity} = \frac{\text{Weight in Air}}{(\text{Weight in Air—Weight in Water})} = \frac{X}{X-Y}$$

Weighing apparatus is not always available; therefore, your work with specific gravity will be an estimate. Most common minerals have a specific gravity of 2.5 to 3.0. Minerals less than 2.5 feel "light" and those greater than 3.0 feel "heavy."

(I) Other Properties—Some minerals *feel* smooth, greasy, soapy, or rough. If this feel is distinctive, it can be a useful characteristic for recognizing the mineral. A few minerals are *attracted to a magnet*. Magnetite is the most common example of a mineral with this property. The variety of magnetite called loadstone is itself a magnet, attracting other magnetic minerals to it. Some minerals have a characteristic *odor*, such as kaolin which smells "earthy" when moistened and pyrite which has a sulfur odor when ground to powder.

A few minerals have a distinctive *taste*, for example halite which is salty. Certain uranium-bearing minerals *give off radiations* that can be detected by sensitive instruments like Geiger counters and scintillometers. Other minerals *emit a colorful glowing light* when exposed to fluorescent light (black light). Calcite and some other carbonate-bearing minerals *effervesce* (bubble and foam) because carbon dioxide (CO_2) is released when dilute acid is placed on them. Each special physical or chemical property as those described is an aid to the identification of the mineral which exhibits that property.

Appendix III

DESCRIPTIONS OF SOME COMMON ROCKS

Most rocks can be identified by recognizing one or more of the mineral constituents and observing the texture (as outlined in Geology Experiments 4 and 5 and Appendix II). However, the following will commonly be of considerable help.

(1) If possible find out where the sample was obtained, as this may suggest what it is.

(2) If the sample looks water worn, be alerted for chert.

(3) If the sample was purchased or contains fragile-looking crystals, it may be relatively rare or in a crystal form which is difficult to identify.

(4) Automatically check for hardness, for reaction to dilute hydrochloric acid, and whether or not you can see more than one mineral constituent.

(5) Be sure you have a fresh surface to observe—this generally means be prepared to chip off a piece (except for fancy crystal specimens or specimens which were purchased).

If all else (tests, reference books, etc.) fails, have a geologist examine the sample. It may require microscopic study.

A brief description of some common rocks, which will aid in their identification, follows.

A. IGNEOUS ROCKS

Basalt (*Volcanic or Plutonic*)—Very fine-grained, fairly heavy, black rock. Individual crystals generally too small to be recognized, even with a hand lens. Plutonic varieties and very dense, volcanic varieties may contain cavities (gas vesicles). Basalt has a general hardness of about 6 and dense varieties break with a conchoidal fracture.

Gabbro (*Plutonic*)—Medium- to coarse-grained, dark green to black rock. Contains abundant pyroxene (augite) and feldspar (plagioclase); may have olivine, amphibole (hornblende), hematite, or magnetite.

Granite (*Plutonic*)—Medium- to coarse-grained, light-colored (gray to pink) rock. Contains abundant quartz and felds-

263

par (both plagioclase and orthoclase or microcline). Biotite mica or less commonly amphibole (hornblende) is prominent as the dark mineral.

Obsidian (*Volcanic*)—Transparent to translucent, generally black glass. Obsidian has a conchoidal fracture and a hardness of about 7.

Pumice (*Volcanic*)—Very fine-grained, light-colored (gray to almost white), frothy glass. Pumice is very light in weight (some samples will float in water).

B. SEDIMENTARY ROCKS

Chert—Extremely fine-grained rock, composed of quartz. It occurs in a variety of colors, most commonly gray, tan, or brown. Conchoidal fracture and hardness of 7. Abundant as water-worn pebbles and cobbles. Indians commonly used varieties of chert, flint, and jasper for artifacts.

Conglomerate—Rounded rock fragments ranging in size from pebbles to boulders cemented together.

Limestone—Fine-grained, light-gray to tan or even black-colored rock. Composed of calcium carbonate (calcite) particles precipitated and cemented together. Commonly limestone contains fossils. It has a hardness of 3. Limestone will effervesce (bubble and foam) from contact with dilute hydrochloric acid.

Dolomite—Similar to limestone but does not react to dilute hydrochloric acid unless it is powdered.

Sandstone—Fine to coarse sand grains (mainly quartz) cemented together. Hardness and color of sandstone depend on the nature of the cementing material. Sand layers are generally visible in the rock. Individual grains can usually be obtained by rubbing the sample on a hard surface. It may contain fossils.

Shale—Fine particles (clay) compacted. Usually shale is gray or black in color. Shale splits into thin plates. May contain fossils. Shale is soft and can be easily scratched.

C. METAMORPHIC ROCKS

Gneiss—Banded or streaked appearance, commonly alternating light- and dark-colored mineral bands. Generally medium or coarse grained. Color depends on mineral composition (which may be of wide variety).

Marble—Generally white and coarse crystalline (with sparkly crystal faces). Marble is recrystallized limestone. Thus it has the properties of calcite—hardness of 3 and will effervesce with dilute hydrochloric acid.

Quartzite—Generally white, gray, tan, or other light colors. Quartzite is recrystallized quartz sandstone. Hence, it has the properties of quartz—hardness of 7 and conchoidal fracture.

Schist—Fine- to medium-grained rock with platy minerals (such as chlorite, biotite mica, muscovite mica, talc) common and oriented in flat-lying layers. Color and composition depend on the minerals it contains.

Slate—Extremely fine-grained rock, generally black but may be blue-gray, green, or other colors. Formed by metamorphism of shale, it is characterized by parting along very flat planes (slaty cleavage) and thus can be used for blackboards and slate-shingled roofs.

Appendix IV

PROCEDURES FOR CALIBRATING A
PHOTOGRAPHER'S LIGHT METER

Meteorology Experiment 5 and Astronomy Experiment 12 use a light meter to measure the intensity of reflected light and the dependence of intensity upon distance. Any light meter whose scale is calibrated in foot-candles can be used directly in these experiments. However, you may find it much less expensive to buy photographer's light meters (or to borrow them from parents who are amateur photographers). These meters are calibrated to read lens openings (f stops) and film speeds (ASA) directly. To calibrate them to read light intensity (in arbitrary units), or to check an unknown scale, follow the procedure below.

EQUIPMENT

- A photographer's light meter
- 4 40-watt light bulbs
- 4 light bulb sockets and cords
- Frosty marking tape

PROCEDURE

Put a piece of frosty marking tape over the face of the light meter. Darken the room, so that the needle hand of the meter moves to its minimum reading. Mark this position on the tape and label it "zero."

Now place the four 40-watt light bulbs close together and turn them on. Move the light meter toward and away from the light bulbs until you find the location where the needle hand makes its maximum reading. Leave the light meter in this location for the remainder of the calibrating procedure. Mark the position of the needle hand on the tape and label it "four."

Turn off one of the four light bulbs and label the new position of the needle hand "three." With only two light bulbs

266

burning, label position "two" of the needle hand. Finally, mark and label the position of the needle hand when only one light bulb is burning.

You have now marked off 0, 1, 2, 3, and 4 arbitrary units of light intensity. If you wish, you may subdivide these units by subdividing the spaces between consecutive marks. (This will give accurate subdivisions unless the space between consecutive marks varies greatly. If this happens, you will have to begin again and use more light bulbs.)

Appendix V

RESOURCE MATERIALS IN EARTH SCIENCE

This appendix is a select listing of materials including books, pamphlets, periodicals, films, and other visual aids. It is not intended to be all inclusive but rather to represent some of the better material available. For a more complete listing, the reader is referred to: Pamphlet RS-2, *Selected References for Earth Science Courses,* by W. H. Matthews III, ESCP Reference Series, 1964; *Oceanography in Print: A Selected List of Educational Resources,* compiled by Lynn Forbes, 1968; and other references cited in the following materials:

GEOLOGY

BOOKS—RESOURCE MATERIALS:

Dana's Manual of Mineralogy, revised by Cornelius Hurlbut, Jr. 17th ed., John Wiley and Sons, Inc., N. Y., 1959.

Dictionary of Geological Terms, by the American Geological Institute. Abridged ed., Doubleday, Garden City, N. Y., 1962.

Earthquakes and Earth Structure, by John H. Hodgson. Prentice-Hall, Inc., Englewood Cliffs, N. J., 1964.

Elements of Geology, by James H. Zumberge. 2d ed., John Wiley and Sons, Inc., N. Y., 1963.

Essentials of Earth History: An Introduction to Historical Geology, by William L. Stokes. 2d ed., Prentice-Hall, Inc., Englewood Cliffs, N. J., 1965.

Geology Illustrated, by John S. Shelton. W. H. Freeman and Company, San Francisco, 1966.

Physical Geology, by Chester R. Longwell and others. John Wiley and Sons, Inc., N. Y., 1969.

Principles of Geology, by James Gilluly and others. 2d ed., W. H. Freeman and Company, San Francisco, 1959.

Search for the Past; an Introduction to Paleontology, by James R. Beerbower. 2d ed., Prentice-Hall, Inc., Englewood Cliffs, N. J., 1968.

BOOKS—GENERAL READING:

A Guide to the National Parks: Their Landscape and Geology, by William H. Matthews III. Natural History Press, N. Y., 1968. Two-volume set, Vol. I–Western Parks, Vol. II–Eastern Parks.

Crystals and Crystal Growing, by Alan Holden and Phylis Singer, Doubleday and Company, Inc., N. Y. (Science Study Series), 1960.

Fossils: An Introduction to Prehistoric Life, by William H. Matthews III. Barnes and Noble, Inc., N. Y., 1962.

Rocks and Minerals, by Richard M. Pearl. Barnes and Noble, Inc., N. Y., 1956.

Rocks and Minerals, by Herbert S. Zim and Paul Shafer. Golden Press, N. Y., 1957.

Speleology: The Study of Caves, by George W. Moore and G. Nicholas. D. C. Heath and Company, Boston, 1964.

The World of Ice, by James L. Dyson. Alfred A. Knopf, N. Y., 1962.

Volcanoes: In History, In Theory, In Eruption, by Fred M. Bullard. University of Texas Press, Austin, 1962.

PAMPHLETS:

"Chemistry and the Solid Earth," *Chemical and Engineering News,* v. 45, October 2, 1967, Washington, D. C. (C & EN Special Report).

Chemistry in Introductory Geology, by W. D. Keller, 4th ed., Lucas Brothers Publishers, Columbia, Missouri, 1969.

Present Problems About the Past, by W. Auffenberg, BSCS Pamphlet 6, D. C. Heath and Company, Boston, Mass., 1963.

The Earth and Its Story: Geology for Young Scientists, by Lou W. Page, American Education Publications, Columbus, Ohio, 1961.

"The Interior of the Earth: An Elementary Description," by Eugene C. Robertson, *U. S. Geological Survey Circular* 352, Washington, D. C., 1966.

PERIODICALS:

Gems and Minerals, Monthly journal published by Gemac Corporation, 1791 Capri Avenue, Mentone, California.

Geotimes, published ten times per year by the American Geological Institute, 2201 M Street, NW, Washington, D. C.

Journal of Geological Education, published five times per year by the National Association of Geology Teachers. (Membership to persons interested in geological education: M. B. Rosalsky, Secretary–NAGT, Department of Geology, City College of New York, N. Y. 10031.)

FILMS:

An Approach to the Prediction of Earthquakes. 28 minutes, color. Iwanami Productions Incorporated, Tokyo, Japan.

Erosion—Leveling the Land. 14 minutes, color. American Geological Institute–Encyclopaedia Britannica Educational Corporation, Chicago, Illinois.

Evidence for the Ice Age. 18 minutes, color. American Geological Institute–Encyclopaedia Britannica Education Corporation, Chicago, Illinois.

Fossils: Clues to Prehistoric Times. 10 minutes, color. Coronet Films, Inc., Chicago, Illinois.

How Solid Is Rock? 25 minutes, color. (Earth Science Curriculum Project Film) American Geological Institute–Encyclopaedia Britannica Educational Corporation, Chicago, Illinois.

Rocks That Form on the Earth's Surface. 16 minutes, color. American Geological Institute–Encyclopaedia Britannica Educational Corporation, Chicago, Illinois.

Rocks That Originate Underground. 23 minutes, color. American Geological Institute–Encyclopaedia Britannica Educational Corporation, Chicago, Illinois.

Why Do We Still Have Mountains? 20 minutes, color. American Geological Institute–Encyclopaedia Britannica Educational Corporation, Chicago, Illinois.

MAPS AND PHOTOGRAPHS:

Geologic Map Portfolios, published by Williams & Heintz Map Corporation, 8119 Central Avenue, Washington, D. C. 20027. Portfolio No. 1: A Laboratory Study of Geologic Maps and Sections (1965), compiled and edited by Louis W. Currier. Portfolio No. 2: A Laboratory Study for Historical Geology (1969), compiled and edited by Forbes Robertson.

Topographic Maps of the United States, published by the U. S. Geological Survey. Standard topographic maps are usually published in 7½- and 15-minute quadrangles and are sold for 50 cents per copy. State topographic mapping indexes for each state are free on request. Each state index shows

270

the areas mapped and gives a list of local dealers who sell topographic maps. Maps and state indexes for all states east of the Mississippi River and including Puerto Rico and the Virgin Islands are available from: **U. S. Geological Survey, Distribution Section, 1200 South Eads Street, Arlington, Va. 22202;** for states west of the Mississippi River and including Guam and American Samoa from: **U. S. Geological Survey, Distribution Section, Federal Center, Denver, Colo. 80225.**

A brochure entitled *Topographic Maps: Descriptive Folder* explaining aspects of topographic map reading, map-making procedures, and availability of maps is available free upon request to the U. S. Geological Survey.

Topographic Maps Illustrating Specific Physiographic Features are available in 100-map and 25-map sets (costs are $30 and $7.50 respectively) from the U. S. Geological Survey. Each map in these sets was chosen because it illustrates a physiographic feature very well. Aerial photographs are available for each of the map areas.

A Descriptive Catalogue of Selected Aerial Photographs of Geologic Features in the United States, U. S. Geological Survey Professional Paper 590, 1969 ($2.25). Available from Superintendent of Documents, Government Printing Office, Washington, D. C. 20402.

OCEANOGRAPHY

BOOKS—RESOURCE MATERIALS:

An Introduction to Marine Geology, by M. J. Keen, Pergamon Publishing Co., Elmsford, N. Y., 1968.

Descriptive Physical Oceanography, by George L. Pickard, Pergamon Publishing Co., Elmsford, N. Y., 1966.

General Oceanography, by Gunter Dietrich, Interscience Publishers, N. Y., 1963.

Oceanography, by M. Grant Gross, Charles E. Merrill Books, Inc., Columbus, Ohio, 1967.

Submarine Geology, by Francis P. Shepard, 2d ed., Harper and Row, Publishers, N. Y., 1963.

The Earth Beneath the Sea, by Francis P. Shepard, Johns Hopkins Press, Baltimore, Md., 1959.

271

The Oceans: Their Physics, Chemistry, and General Biology, by H. U. Sverdrup, Martin W. Johnson. and Richard H. Fleming, Prentice-Hall, Inc., N. J., 1942.

The Sea off Southern California, by K. O. Emery, John Wiley and Sons, Inc., N. Y., 1960.

BOOKS—GENERAL READING:

Waves and Beaches: The Dynamics of the Ocean Surface, by Willard Bascom. Doubleday, Garden City, N. Y. (Science Study Series), 1964.

Riches of the Sea: The New Science of Oceanography, by Norman Carlisle (Advances in Science Series), Sterling Publishing Co., Inc., New York, 1967.

The Sea Around Us, by Rachel L. Carson, rev. ed., Oxford University Press, Inc., New York, 1961

Searchers of the Sea: Pioneers in Oceanography, by Charles M. Daugherty, The Viking Press, Inc., New York, 1961.

Water: The Mirror of Science, by Kenneth S. Davis and John A. Day, Doubleday and Company, Inc., N. Y. (Science Study Series), 1961.

Seas, Maps, and Men: An Atlas-History of Man's Exploration of the Oceans, G. E. R. Deacon, General Editor, Doubleday and Company, Inc., N. Y., 1962.

The Sea, by Leonard Engel, Time, Inc. (Life Nature Library), N. Y., 1961.

The Global Sea, by Harris B. Stewart, Jr., D. Van Nostrand Co., Inc., Princeton, N. J., 1963.

PAMPHLETS:

A Reader's Guide to Oceanography, by Jan Hahn, Woods Hole Oceanographic Institution, Woods Hole, Mass., 1961.

Oceanography, by John C. Weihaupt, United States Armed Forces Institute, Madison, Wisconsin, 1966.

Oceanography in Print: A Selected List of Educational Resources, by Lynn Forbes, Compiler, Oceanography Education Center, Falmouth, Mass., 1968. (A very useful pamphlet listing books, films, and periodicals in oceanography.)

Science and the Sea, by U. S. Naval Oceanographic Office, Washington, D. C., 1967.

Secrets of the Sea: Oceanography for Young Scientists, by Howard J. Pincus, American Education Publications, Columbus, Ohio, 1960.

PERIODICALS:

Ocean Industry, published monthly by the Gulf Publishing Company, P. O. Box 2068, Houston, Texas.

Oceanology International, published seven times a year by Industrial Research Publications Co., Beverly Shores, Indiana.

Oceans, published monthly, P. O. Box 2406, La Jolla, California.

Sea Frontiers, published monthly by International Oceanographic Foundation, 1 Rickenbacker Causeway, Virginia Key, Miami, Florida.

Sea Secrets, published 11 times per year by International Oceanographic Foundation, 1 Rickenbacker Causeway, Virginia Key, Miami, Florida.

FILMS:

Challenge of the Oceans. 27 minutes, color. McGraw-Hill Text-Films, N. Y. (Planet Earth Film Series).

Conquering the Sea. 25 minutes, color. McGraw-Hill Text-Films, N. Y. (The 21st Century Series).

Ocean Basins. 16 minutes, color. Lamont Laboratory, Columbia University, N. Y. (Marine Science Film Series).

Restless Sea. 54 minutes, color. Bell Telephone Company, Regional Bell Telephone Company, Business Office.

Science of the Sea. 19 minutes, color. International Film Bureau, Chicago, Illinois.

The Deep Frontier. 25 minutes, color. McGraw-Hill Text-Films, N. Y. (The 21st Century Series).

The Beach—A River of Sand. 20 minutes, color. American Geological Institute–Encyclopaedia Britannica Educational Corporation, Chicago, Illinois.

Waves on Water. 16 minutes, color. American Geological Institute–Encyclopaedia Britannica Educational Corporation, Chicago, Illinois.

METEOROLOGY

BOOKS—RESOURCE MATERIALS:

An Introduction to Climate, by G. T. Trewartha. McGraw-Hill Book Company, Inc., N. Y., 1954.

273

General Meteorology, by Horace R. Byers. 3d ed., McGraw-Hill Book Company, Inc., N. Y., 1959.

Glossary of Meteorology, by Ralph E. Huschke, editor. American Meteorological Society, Boston, Mass., 1959.

International Cloud Atlas, by World Meteorological Organization, Geneva, Switzerland, 1956 (official atlas of cloud types).

The Ocean of Air, by David J. Blumenstock. Rutgers University Press, New Brunswick, N. J., 1959.

Weather and Man, by Hans H. Neuberger and F. Briscoe Stephens. Prentice-Hall, Inc., Englewood Cliffs, N. J., 1948.

Weathercasting, by Charles Laird and Ruth Laird. Prentice-Hall, Inc., Englewood Cliffs, N. J., 1955.

Weather Elements, by Thomas A. Blair and Robert C. Fite. 5th ed., Prentice-Hall, Inc., Englewood Cliffs, N. J., 1965.

BOOKS—GENERAL READING:

Cloud Physics and Cloud Seeding, by Louis J. Battan. Doubleday and Company, Inc., N. Y. (Science Study Series), 1962.

Cloud Study, by Frank H. Ludlam and Richard S. Scorer. The Macmillan Company, N. Y., 1958.

Everyday Weather and How It Works, by Herman Schneider. rev. ed., McGraw-Hill Book Company, Inc., N. Y., 1961.

Radar Observes the Weather, by Louis J. Battan, Doubleday and Company, Inc., N. Y. (Science Study Series), 1962.

The Restless Atmosphere, by Fredrick K. Hare. Harper and Row, Publishers, N. Y., 1961.

Water, by Luna B. Leopold, Kenneth S. Davis, and the Editors of Life. Time–Life Inc., N. Y. (Life Science Library), 1966.

Weather, by Philip D. Thompson, Robert O'Brien and the Editors of Life. Time–Life Inc., N. Y. (Life Science Library), 1965.

Weather. A Guide to Phenomena and Forecasts, by Paul E. Lehr, R. Will Burnett, and Herbert S. Zim. Golden Press, Inc., N. Y., 1957.

PAMPHLETS:

Little Climates, by Verne M. Rockcastle. Cornell University Press, Ithaca, N. Y., 1961.

Manual of Lecture Demonstrations, Laboratory Experiments, and Observational Equipment for Teaching Elementary Meteorology in Schools and Colleges, by Hans Neuberger and George Nicholas. Pennsylvania State University Press, University Park, Pa., 1962.

The Weather Workbook, by Fred W. Decker. rev. ed., The

Weather Workbook Co., 827 N. 31st St., Corvallis, Oregon, 1966.

Weather and You, by Willard L. Leeds. American Education Publications, Columbus, Ohio, 1962.

Weather Forecasting, by U. S. Weather Bureau. U. S. Government Printing Office, Washington, D. C., 1952.

PERIODICALS:

Weather, published monthly by The Royal Meteorological Society, 49 Cromwell Rd., London, S.W. 4, England.

Weatherwise, published bimonthly by The American Meteorological Society, 45 Beacon St., Boston, Mass.

FILMS:

Reading Weather Maps. 13½ minutes, color. Coronet Films, Inc., Chicago, Illinois.

The Formation of Raindrops. 26 minutes, color. Modern Learning Aids, New York, N. Y. (American Meteorological Society Production).

The Inconstant Air. 27 minutes, color. McGraw-Hill Text-Films, N. Y. (Planet Earth Film Series).

The Unchained Goddess. 57 minutes, color. Bell Telephone Company, Regional Bell Telephone Company Business Office.

Weather: Understanding Storms. 11 minutes, black and white. Coronet Films, Inc., Chicago, Illinois.

Weather: Why It Changes. 11 minutes, color. Coronet Films, Inc., Chicago, Illinois.

What Makes Clouds? 19 minutes, color. American Geological Institute–Encyclopaedia Britannica Educational Corporation, Chicago, Illinois.

What Makes the Wind Blow? 16 minutes, color. American Geological Institute–Encyclopaedia Britannica Educational Corporation, Chicago, Illinois.

ASTRONOMY

BOOKS—RESOURCE MATERIALS:

A Fundamental Survey of the Moon, by Ralph B. Baldwin. McGraw-Hill Book Company, Inc., N. Y., 1965.

Astronomy, by Robert H. Baker. 7th ed., D. Van Nostrand Company, Inc., Princeton, N. J., 1959.

Astronomy, by the Elementary School Science Project, Univ. of Illinois Press, Urbana, Illinois, 1962-65 (six volumes).

Exploration of the Universe, by George O. Abell. Holt, Rinehart and Winston, Inc., N. Y., 1964.

Introduction to Astronomy, by Dean B. McLaughlin. Houghton Mifflin Company, Boston, Mass., 1961.

Principles of Astronomy, by Stanley P. Wyatt. Allyn and Bacon, Inc., Boston, Mass., 1964.

Sourcebook on the Space Sciences, by Samuel Glasstone. D. Van Nostrand Company, Inc., Princeton, N. J., 1965.

Tools of the Astronomer, by Gerhard R. Miczaika and William M. Sinton. Harvard University Press, Cambridge, Mass., 1961.

BOOKS—GENERAL READING:

And There Was Light: The Discovery of the Universe, by Rudolf Thiel. Alfred A. Knopf, Inc., N. Y., 1957.

Exploring the Universe, by Roy A. Gallant. Doubleday and Company, Inc., Garden City, N. Y., 1956.

Naked-Eye Astronomy, by Patrick Moore. W. W. Norton and Company, Inc., N. Y., 1966.

Planets, by Carl Sagan and H. N. Leonard. Time–Life Inc., N. Y. (Life Science Library), 1968.

Seeing the Earth from Space, by Irving Adler. The John Day Company, Inc. (Signet), N. Y., 1962.

The Moon: Earth's Natural Satellite, by Franklyn M. Branley. Thomas Y. Crowell Company, N. Y., 1960.

The Universe, by David Bergamini and the Editors of Life, Time–Life Inc., N. Y. (Life Nature Series), 1962.

The Universe Around Us, by Sir James Jeans. 4th ed., Cambridge University Press, N. Y., 1960.

PAMPHLETS:

Astronomy, by Clyde Fisher. (Boy Scout Merit Badge, Pamphlet No. 3303), Boy Scouts of America, New Brunswick, N. J., 1962.

Astronomy Highlights, series of eight booklets with Thomas D. Nicholson, Series Editor. The Natural History Press, Garden City, N. Y., 1964. Titles are:
"Apollo and the Moon" by Franklyn M. Branley.
"Birth and Death of the Stars" by Kenneth L. Franklin.

"Captives of the Sun" by James S. Pickering.

"Design of the Universe" by S. I. Gale.

"Man in Space" by Fred C. Hess.

"Space Age Astronomy" by Kenneth L. Franklin.

"The Sun in Action" by Thomas D. Nicholson.

"Time and the Stars" by Joseph M. Chamberlain.

A Dictionary of Astronautics, by J. A. Nayler. Hart Publishing Company, N. Y., 1964.

Exploring the Universe, by Allen Hynek. American Education Publications, Columbus, Ohio, 1961.

PERIODICALS:

Popular Astronomy, published bimonthly by Sky Map Publishers, Inc., 111 South Meramec, St. Louis, Missouri.

Leaflets of the Astronomical Society of the Pacific, published bimonthly by the Astronomical Society of the Pacific, c/o California Academy of Sciences, Golden Gate Park, San Francisco, California.

Sky and Telescope, published monthly by Sky Publishing Company, Harvard College Observatory, 49-50-51 Bay State Road, Cambridge, Mass.

FILMS:

A Radio View of the Universe. 28½ minutes, color. Modern Learning Aids (1212 Avenue of the Americas, New York, N. Y.).

A Trip to the Moon. 16 minutes, color. American Geological Institute–Encyclopaedia Britannica Educational Corporation, Chicago, Illinois.

Causes of the Seasons. 11 minutes, color. Coronet Films, Inc., Chicago, Illinois.

Exploring the Universe. 11 minutes, black and white. Encyclopaedia Britannica Educational Corporation, Chicago, Illinois.

Frames of Reference. 28 minutes, black and white. Modern Learning Aids (PSSC film), New York, N. Y.

How We Know the Earth Moves. 10 minutes, color. Film Associates of California, Los Angeles, California.

It's About Time. 50 minutes, color. Bell Telephone Company, Regional Bell Telephone Company Business Office.

The Mystery of Stonehenge. 57 minutes, color. McGraw-Hill Text-Films, N. Y.

EARTH SCIENCE

BOOKS—RESOURCE MATERIALS:

Earth and Space Science, by C. Wroe Wolfe, Louis J. Battan, Richard H. Fleming, Gerald S. Hawkins, and Helen Skornik, D. C. Heath and Company, Boston, Mass., 1966.

Earth Science, by Richard J. Ordway, D. Van Nostrand Company, Inc., Princeton, N. J., 1966.

Earth Science: The World We Live In, by Samuel N. Namowitz and Donald B. Stone, 3rd ed., D. Van Nostrand Company, Inc., Princeton, N. J., 1965.

Fundamentals of Earth Science, by Henry D. Thompson, 2nd ed., Appleton-Century-Crofts, Inc., N. Y., 1960.

Geology and Earth Sciences Sourcebook for Elementary and Secondary Schools, Robert L. Heller, Editor (program developed by the American Geological Institute), Holt, Rinehart and Winston, Inc., New York, 1962.

Investigating the Earth, by Earth Science Curriculum Project (a project of the American Geological Institute), Houghton Mifflin Company, Boston, Mass., 1967.

Modern Earth Science, by William J. Ramsey and Raymond A. Burckley, 2nd ed., Holt, Rinehart and Winston, Inc., N. Y., 1965.

Physical Science Library, Charles E. Merrill Publishing Co., Columbus, Ohio.
  Astronomy, by E. G. Ebbighausen
  Geology, by Robert J. Foster
  Meteorology, by Albert Miller
  Oceanography, by M. Grant Gross

The Earth Sciences, by Arthur N. Strahler, Harper and Row, Publishers, N. Y., 1963.

Time, Space, and Matter: Investigating the Physical World (Secondary School Science Project, Princeton, N. J.) published by Webster Division, McGraw-Hill Book Company, Inc., St. Louis, Mo.

BOOKS—GENERAL READING AND CAREER MATERIALS:

Aerial Stereo Photographs, by Harold R. Wanless, T. N. Hubbard Scientific Co., Northbrook, Ill., 1965, 92 plates.

Inquiry Techniques for Teaching Science, by William D. Romey, Prentice-Hall, Inc., N. J., 1968.

Stereo Atlas, by Earth Science Curriculum Project (a project of the American Geological Institute), T. N. Hubbard Scientific Co., Northbrook, Ill., 1968. 48 plates (also published by Raytheon Education Co., 186 Third Avenue, Waltham, Mass).

The Earth, by Arthur Beiser and the editors of Time–Life Inc., N. Y. (Life Nature Library), 1962.

Your Future in Geology, by Joseph L. Weitz, Richards Rosen Press, Inc., N. Y., 1966.

Your Future in Meteorology, by Frederick A. Berry and Sidney R. Frank, Richards Rosen Press, Inc., N. Y., 1962.

Your Future in Oceanography, by Norman H. Gaber, Richards Rosen Press, Inc., N. Y., 1967.

PAMPHLETS:

Chronicle Occupational Briefs, published by Chronicle Guidance Publishers, Inc., Moravia, N. Y. (pamphlets on individual careers in astronomy, geology, geography, meteorology, oceanography).

Earth Science Curriculum Project Reference Series edited by William H. Matthews III (program developed by the American Geological Institute), published by Prentice-Hall, Inc., N. J.

RS-1 "Sources of Earth Science Information" by William H. Matthews III, 1964.

RS-2 "Selected References for Earth Science Courses" by William H. Matthews III, 1964.

RS-3 "Selected Earth Science Films" by Wakefield Dort, Jr., 1964.

RS-4 "Selected Maps and Earth Science Publications" (for the states and provinces of North America) by William H. Matthews III, 1965.

RS-5 "Free Materials for Earth Science Teachers" by William H. Matthews III and Rolland B. Bartholomew, 1965.

RS-6 "Planetariums, Observatories and Earth Science Exhibits" by William H. Matthews III, 1965.

RS-7 "Topographic Maps and How to Use Them" by Malcolm P. Weiss, 1967.

RS-8 "Basic Data and Water Budget Computation" (for selected cities in North America) by Douglas B. Carter, 1967.

RS-9 "Selected Guides for Geologic Field Study in Canada

and the United States of America" by Donald
H. Lokke, 1967.

Scientific American Offprints reprinted by W. H. Freeman and
Company, 660 Market Street, San Francisco, California
94104. (Reprints of over 100 articles on aspects of earth
science published in *Scientific American.*)

The Sphere of the Geological Scientist, by Chalmer J. Roy, American Geological Institute, Washington, D. C., 1965.

PERIODICALS:

National Geographic, published monthly by the National Geographic Society, 17th and M Streets NW, Washington, D. C.

Natural History, published monthly October through May, bi-
monthly June to September for the American Museum of
Natural History by the Natural History Press, Central Park
West at 79th Street, N. Y.

Science, published weekly by the American Association for the
Advancement of Science, 1515 Mass. Ave., NW, Washington, D. C.

Science and Children, published monthly September through December and February through May by the National Science
Teachers Association, 1201 Sixteenth Street NW, Washington, D. C.

Science Digest, published monthly by the Hearst Corporation,
Box 654, N. Y.

Science News, published weekly by Science Service, Inc. (an
Institution for the Popularization of Science), 1719 N Street,
NW, Washington, D. C.

Scientific American, published monthly by Scientific American
Inc. 415 Madison Avenue, N. Y.

The Science Teacher, published monthly September through May
by the National Science Teachers Association, 1201 Sixteenth Street, NW, Washington, D. C.

VISUAL AIDS:

Curriculum Color Prints, Instructional Aids Incorporated, Box
191, Mankato, Minnesota 56001. (Includes earth history,
earth science, and oceanography.)

Earth Science Transparency Series, Hammond Incorporated, 515
Valley Street, Maplewood, N. J. 07040. (Set of 24 transparencies in astronomy, geology/physical geography,
meteorology, and oceanography, 1966.)

Earth Sciences Transparency Masters, The Keuffel and Esser

Company, 20 Whippany Road, Morristown, N. J. 07960. (Book of 102 transparency masters in astronomy, geology, meteorology, oceanography, and related sciences.)

Filmstrips in Earth Science, McGraw-Hill Text-Films, 330 W. 42nd St., N. Y. 10018. (The Story of the Universe Filmstrip Series.)

Filmstrips and Study Prints, Society for Visual Education, Inc., 1345 Diversey Parkway, Chicago, Ill. 60614. (Filmstrips and study prints in earth science and space science.)

Projectuals, Visucom, Chicopee, Mass. 01020. (Projectuals or transparencies in astronomy, geology, and meteorology.)

Toward Inquiry (Teacher Preparation Film), Encyclopaedia Britannica Educational Corporation, 1150 Wilmette Avenue, Wilmette, Ill. 60091. (Black and white film, 20 minutes, prepared in cooperation with the Earth Science Curriculum Project, 1967.)

Transparencies, The Instructo Corporation, Paoli, Pa. 19301. (Series of teaching transparencies in earth and space science.)

COMPANIES WITH EARTH SCIENCE MATERIALS:

A. J. Nystrom and Company, 3333 Elston Avenue, Chicago, Ill. 60618. (Charts, globes, maps, transparencies, etc. in earth science.)

American Petroleum Institute, 1271 Avenue of the Americas, N. Y. 10020. (Wall charts and pamphlets on earth science distributed without charge to teachers.)

Colorado Geological Industries, Inc., 1244 East Colfax Avenue, Denver, Colorado 80218. (Specialize in rocks and minerals ranging from quantity supplies of common types to rare specimens.)

Damon Engineering, Incorporated, Educational Division. 115 Fourth Avenue, Needham, Mass. 02194. (Wide selection of earth science materials.)

Denoyer-Geppert Company, 5235 Ravenswood Avenue, Chicago, Illinois 60640. (Charts, globes, maps, and resource units for use in earth science courses.)

Edmund Scientific Company, 402 Edscorp Building, Barrington, N. J. 08007. (Extensive line of science equipment, including a variety of earth science materials.)

Hubbard Scientific Company, Box 105, Northbrook, Ill. 60062. (Extensive earth science materials including film loops, equipment for laboratory investigations, transparencies, models, etc.)

Raytheon Education Company, 285 Columbus Avenue, Boston, Mass. 02116. (Wide selection of earth science materials.)

Sargent-Welch Scientific Company, 7300 N. Linder Avenue, Skokie, Illinois 60067. (Wide selection of earth science materials.)

Science Kit Incorporated, 2299 Military Road, Tonawanda, N. Y. 14150. (Earth science materials including film loops, equipment for laboratory investigations, and replicas of fossils.)

Science Research Associates, Inc., 259 East Erie Street, Chicago, Illinois 60611. (Inquiry development program in earth science including reference material, equipment kits, and film loops.)

Ward's Natural History Establishment, Inc., Box 1712, Rochester, N. Y. 14603. (Extensive earth science materials including filmstrips, color slides, and fossils.)

Index